ADVANCES IN OXYTOCIN RESEARCH

Pharmacology - Research, Safety Testing and Regulation

Additional books in this series can be found on Nova's website under the Series tab.

Additional e-books in this series can be found on Nova's website under the e-book tab.

PHARMACOLOGY - RESEARCH, SAFETY TESTING AND REGULATION

ADVANCES IN OXYTOCIN RESEARCH

EVAN ANDERSON
EDITOR

New York

Library of Congress Cataloging-in-Publication Data

LCCN: 2015953818

ISBN: 978-1-63483-991-4

Published by Nova Science Publishers, Inc. † New York

CONTENTS

PREFACE

Oxytocin (OT), a hormone with a well-known function in childbirth, breast milk secretion, formation of the mother-infant bonding process and social empathy, has recently been revisited from a systemic perspective to uncover possible links with or etiological characteristics of chronic noncommunicable diseases (NCDs). Chapter one aims to provide an overview of the association of OT and its genetic polymorphisms with the most prevalent NCDs in the world population, seeking to integrate it to the thrifty phenotype hypothesis throughout the ontogenetic process. Chapter two examines the experimental evidence accumulated over the last few decades which indicates that oxytocin can play a beneficial role in cardiovascular control. Chapter three studies recent evidence that demonstrates intranasal inhalation of oxytocin improves face-processing. The final chapter contributes to a better understanding of the mechanisms involved in the age-related changes in body composition, sexual, and feeding behavior, in which oxytocin and cortisol probably play a relevant role.

Chapter 1 – Chronic noncommunicable diseases (NCDs) are a major public health problem due to their high prevalence. In this sense, different emerging risk factors that share interconnected metabolic pathways are of extreme interest in both preventive and therapeutic approaches to NCDs. Oxytocin (OT), a hormone with a well-known function in childbirth, breast milk secretion, formation of the mother-infant bonding process and social empathy, has recently been revisited from a systemic perspective to uncover possible links with or etiological characteristics of NCDs. Various lines of investigation have shown that OT also plays a role as a neurotransmitter and neuromodulator in the central nervous system. Oxytocin in this system has been linked to modulation of maternal behavior and social memory, and to increased empathy, in addition to being related to psychiatric disorders. The scientific literature further indicates that OT synthesized in the heart and vascular smooth muscles plays a key role in the secretion of natriuretic hormone, helping to decrease blood pressure and blood volume, heart rate, vasodilation, and cardiovascular homeostasis and regeneration. These central effects of OT are mediated by oxytocin receptors (OXTR), which are found in various tissues (reproductive system, mammary glands, pancreas, kidney, adipocytes, macrophages and cardiovascular system). It is postulated that OT may be negatively impacted, based on the thrifty phenotype hypothesis proposed by Hales and Barker, suggesting that embryonic or fetal development is sensitive to the intrauterine nutritional environment. However, although a poor intrauterine nutritional environment can damage the formation and consolidation of the oxytocinergic system of the embryo or fetus, it is also believed that a lack of affection and maternal well-being may similarly cause harm to the individual and can be considered another strong component of the thrifty phenotype. Even if stress and gestational depression interfere negatively in the development of key fetal organs, OT is responsible in the post-natal period for the modulation of these harmful effects on survival and adaptation to the environment. Yet, despite increasing the chances of postnatal survival, these maternal factors may alter fetal metabolism, which can then trigger NCDs in adult life. In addition, studies have suggested that the gene encoding the oxytocin receptor (OXTR) is associated with NCDs. Therefore, this review aims to provide an overview of the association of OT and its genetic polymorphisms with the most prevalent NCDs in the world population, seeking to integrate it to the thrifty phenotype hypothesis throughout the ontogenetic process.

Chapter 2 – Experimental evidence accumulated over the last few decades indicates that the nonapeptide hormone oxytocin, which is synthesized both centrally in the hypothalamus, and peripherally in the heart and the vasculature, can play a beneficial role in cardiovascular control. In the periphery, oxytocin has been found to be a growth and differentiation factor in the developing heart that can also promote regeneration and cardiac healing in ischemia. Both oxytocin released in to the blood and oxytocin derived from the heart can increase the release of atrial natriuretic peptide and hence affect hydromineral homeostasis. Quiescence of oxytocin neurons in the hypothalamic paraventicular nucleus, and the concomitant hyperactivity of vasopressin neurons in the paraventicular nucleus, has been documented to underlie osmotic adjustment to pregnancy. In most vascular beds, oxytocin was found to exert weak vasoconstriction by the stimulation of vasopressin V1a receptors. In contrast, the vasoconstriction of the umbilical artery at term by oxytocin is strong. In the vascular beds of the lungs and the brain, oxytocin can produce vasodilatation via NO production. In the central nervous system, oxytocin acts in a paracrine and autocrine fashion, as well as a classical neurotransmitter impinging either directly in the brainstem on neurogenic cardio-respiratory control, or indirectly by the modulation of behavior and emotions. In the paraventicular nucleus, oxytocin receptors were reported to increase baro-receptor reflex sensitivity, reduce blood pressure short-term variability and buffer the cardiovascular response to stress. Thus, recent data suggests multiple roles for oxytocin in cardiovascular regulation. Further research may reveal therapeutic potential.

Chapter 3 – Oxytocin is a nonapeptide that plays a fundamental role in social cognition, and recent evidence demonstrates that intranasal inhalation of the hormone improves face-processing. For example, a number of studies have found that people are significantly better at recognising facial expressions of emotion following oxytocin inhalation; and there is some evidence that oxytocin improves facial identity memory. However, the benefits of oxytocin for facial identity recognition are somewhat inconsistent, with some studies finding effects under limited circumstances (e.g., only when the faces show certain emotional expressions, or only amongst certain groups of participants), and others finding selective, null or negative effects. Notably, some evidence suggests that oxytocin is not always facilitative, raising the possibility that it may only improve face recognition performance under certain conditions. Currently, the mechanisms underpinning the link between oxytocin and improved face-processing are unclear. Oxytocin appears to improve face encoding (as opposed to retrieval of a face from memory), but it is unclear whether this effect arises due to changes in face perception or later encoding or memory consolidation processes, nor why this effect occurs. It is possible that oxytocin acts upon the core face-processing system, and some neuroimaging evidence supports this possibility. Alternatively, oxytocin may affect face-processing indirectly by modulating social salience, approach-withdrawal motivation, or general levels of anxiety. This commentary will summarise the available literature on this issue, attempting to clarify when oxytocin does and does not benefit face recognition, and the mechanisms that underpin this effect.

Chapter 4 – The increase in the elderly population has drawn attention to diseases whose incidence is high in this age group, such as obesity. A common characteristic linked to the aging process is the modification of body composition, mainly increased fat mass. Age-related obesity has a multifactorial origin, mainly involving the interaction of genetic, hormonal and environmental factors. Recent literature has suggested that the pathophysiology of obesity is related to not only neuroendocrine disorders involving the control of appetite and satiety but also neuroendocrine routes that are common to psychiatric (depression and anxiety) and behavioral disorders. In this sense, the hormone oxytocin can be the trigger for the behavioral changes associated with aging, sexual and feeding behavior, as these behaviors share common reward mechanisms involved in the stimulation and feeling of pleasure. Therefore, it is important to emphasize that with aging, disorders may occur in the synthesis and secretion of oxytocin, which may play an etiologic role in not only sexual behavior but also in food and energy metabolism, thus favoring the onset of obesity. Oxytocin is synthesized in the paraventricular and supraoptic nuclei of the hypothalamus that acts in the hippocampus and the amygdala. Therefore, oxytocin signaling can influence the release of other hormones such as cortisol. Because cortisol also acts in the hippocampus and the amygdala, these two hormones can interact. However, information about this correlation between oxytocin and cortisol with respect to sexual and eating behaviors and obesity in the elderly are scarce and controversial. Oxytocin seems to be a component of the intricate network of neurophysiological processes underlying sexual responses. The literature has shown that oxytocin plays various roles in sexual responses, such as uterine contraction, ejaculation, orgasm and feeling of pleasure. In this context, one cannot rule out that dietary intake can also provide a feeling of pleasure and an anti-stress effect similar to the sexual act and that cortisol and oxytocin exert effects on the central nervous system. Therefore, unifying the information regarding the functions performed by oxytocin, it seems plausible that there is an association among aging, obesity (via feeding behavior), and sexual behavior and oxytocin levels. It is within this context that this manuscript aims to contribute to a better understanding of the mechanisms involved in the age-related changes in body composition, sexual, and feeding behavior, in which oxytocin and cortisol probably play a relevant role.

Chapter 1

OXYTOCIN AND CHRONIC NONCOMMUNICABLE DISEASES: ON OVERVIEW OF OXYTOCIN RECEPTOR GENE (OXTR) POLYMORPHISMS AND THE THRIFTY PHENOTYPE HYPOTHESIS

Camila Bittencourt Jacondino[1,*]
and Maria Gabriela Valle Gottlieb[2]

[1]Nurse, PhD student Biomedical Gerontology by the Pontifical Catholic
University of Rio Grande do Sul, Porto Alegre, Brazil
[2]Biogerontologist, Researcher and professor of the Institute of Geriatrics
and Gerontology (IGG), PUCRS, Porto Alegre, Brazil

ABSTRACT

Chronic noncommunicable diseases (NCDs) are a major public health problem due to their high prevalence. In this sense, different emerging risk factors that share interconnected metabolic pathways are of extreme interest in both preventive and therapeutic approaches to NCDs. Oxytocin (OT), a hormone with a well-known function in childbirth, breast milk secretion, formation of the mother-infant bonding process and social empathy, has recently been revisited from a systemic perspective to uncover possible links with or etiological characteristics of NCDs.

---

[*] E-mail: camilabjacondino@gmail.com.

Various lines of investigation have shown that OT also plays a role as a neurotransmitter and neuromodulator in the central nervous system. Oxytocin in this system has been linked to modulation of maternal behavior and social memory, and to increased empathy, in addition to being related to psychiatric disorders. The scientific literature further indicates that OT synthesized in the heart and vascular smooth muscles plays a key role in the secretion of natriuretic hormone, helping to decrease blood pressure and blood volume, heart rate, vasodilation, and cardiovascular homeostasis and regeneration. These central effects of OT are mediated by oxytocin receptors (OXTR), which are found in various tissues (reproductive system, mammary glands, pancreas, kidney, adipocytes, macrophages and cardiovascular system). It is postulated that OT may be negatively impacted, based on the thrifty phenotype hypothesis proposed by Hales and Barker, suggesting that embryonic or fetal development is sensitive to the intrauterine nutritional environment. However, although a poor intrauterine nutritional environment can damage the formation and consolidation of the oxytocinergic system of the embryo or fetus, it is also believed that a lack of affection and maternal well-being may similarly cause harm to the individual and can be considered another strong component of the thrifty phenotype. Even if stress and gestational depression interfere negatively in the development of key fetal organs, OT is responsible in the post-natal period for the modulation of these harmful effects on survival and adaptation to the environment. Yet, despite increasing the chances of postnatal survival, these maternal factors may alter fetal metabolism, which can then trigger NCDs in adult life.

In addition, studies have suggested that the gene encoding the oxytocin receptor (OXTR) is associated with NCDs. Therefore, this review aims to provide an overview of the association of OT and its genetic polymorphisms with the most prevalent NCDs in the world population, seeking to integrate it to the thrifty phenotype hypothesis throughout the ontogenetic process.

1. INTRODUCTION

1.1. Oxytocin and Chronic Noncommunicable Diseases (NCDs): Oxytocin May Be Influenced by the Thrifty Phenotype Hypothesis?

Cardiovascular and chronic lung disease, cancer and diabetes mellitus (DM) are the main noncommunicable diseases (NCDs) worldwide, being responsible for the death of three in five people. In addition, mental health problems such as depression and Alzheimer's disease, metabolic disorders such as overweight and obesity, osteoporosis and chronic kidney disease have all increased similarly, creating a serious public health problem. Such findings demonstrate the importance of prevention of these diseases and they pose a constant challenge to managers, researchers and society as a whole, as programs of prevention and treatment are sought [1-6]. In this way, different emerging risk factors that share interconnected metabolic pathways are of extreme interest to both preventive and therapeutic approaches to NCDs.

Oxytocin (OT), a hormone with a well-known function in childbirth, breast milk secretion, formation of the mother-infant bonding process and social empathy, has recently been revisited from a systemic perspective to uncover possible links with or etiological characteristics of NCDs.

An increased understanding of the various roles played by OT has occurred over recent years. Research has shown that, in addition to its already well-known functions, this hormone operates in different tissues and systems, like the cardiovascular and nervous systems. OT plays a key role in the central nervous system (CNS) as a neuropeptide, neurotransmitter and neuromodulator, and changes in this molecule are closely related to NCDs, such as in depression and possibly Alzheimer's disease.

Neuropeptides from the hypothalamus, among them oxytocin, are critical in maintaining energy balance and basal metabolism [7], which are important for the prevention of NCDs and their complications. OT is a nonapeptide hormone produced by the magnocellular and parvocellular neurons of the paraventricular and supraoptic nuclei of the hypothalamus. It is transported to the pituitary, where it is secreted into the bloodstream to act peripherally, such as to stimulate uterine contractions during parturition and milk ejection. OT also acts as a neuromodulator, being a neurohormone released through magnocellular projections that are located in the paraventricular and supraoptic nuclei of the CNS [8]. In its function as a neuromodulator, OT mediates social and reproductive behavior, including mating, pair-bond formation and maternal behavior [9].

The limbic system is considered the center of emotions and is the main expression site (hippocampus and amygdala) of oxytocin receptors, which is one of the reasons for the association of OT with affective disorders, maternal and social behaviors, social memory and increases in empathy [10, 11]. Studies with mammals have revealed that OT is responsible for the construction of behavior, with recognition of conspecifics for partner preference formation for pair-mating, and the expression of certain emotions, such as fear and anxiety [11].

OT levels in blood plasma have been shown to present in high concentrations in both the mother and baby in the first trimester of pregnancy, as well as the first month post childbirth. In the latter case, the high concentrations are due to sensory stimulation (tactile, thermal and olfactory), especially during the breast-feeding process, and it exerts a calming effect on the baby, strengthening the emotional bond between mother and child, and stimulating prosocial behavior in the newborn [12, 13]. A longitudinal study of 18 human newborns showed that infants with higher levels of OT appear to actively seek social interaction [14].

As with all altricial species, mother-child contact in the first months of life is essential for the development and appropriate growth and health of the offspring, as the mother is responsible for the regulation of several physiological responses of the young, for example, heart rate, sleep-wake cycle, and growth hormone (GH) production. Conversely, maternal deprivation causes deleterious effects on offspring development, such as reduced growth, cell differentiation and GH secretion. In addition, studies with rats have demonstrated that separation of the pups from the litter causes persistent changes in the functioning of the hypothalamic-pituitary-adrenal (HPA) axis, as well as behavioral, metabolic and neurochemical alterations [15].

There is a susceptibility linked to the plasticity of the cerebral structures in this phase. The CNS in both rodents and humans is still forming during the first few days after birth [16].

Difficulties in the early social setting can also adversely affect expression of the OT-receptor gene, suggesting that the initial environment directly influences OT concentrations in humans [17]. A study by Fries et al., [18] involving children living in an unfavorable social situation, observed that they had low OT concentrations, as well as little social interaction. This result suggests that early environment can influence the transmission of social ties, as well as behavior. The research highlights that early life is a vital period for the healthy development of humans and that OT is an important regulator of emotional development [19].

Beyond these functions related to neuropsychiatric disorders, OT has now been associated with metabolic homeostasis and cardiovascular, renal and bone tissue diseases.

Investigations have shown that OT receptors (OXTR) are expressed in many organs, such as kidneys, heart, pancreas and adipose and bone tissues, in addition to their relationship with anti-stress mechanisms due to their action on the HPA axis [20, 21].

The nervous system itself, including synthesis and secretion of neurotransmitters and neuromodulators, probably also suffers damage in an inappropriate environment. This can negatively interfere in the construction of a healthy and plastic neuronal network in postnatal life. Furthermore, investigations have shown that other factors, besides nutrition, can affect embryonic and fetal development, such as for example, maternal behavior in relation to stressful situations. The intrauterine environment may influence fetal development, leading to the emergence of possible long-term consequences for child development [22, 23].

In this scenario, a mother who experiences some form of stressful event will have, as a consequence, hyperactivity of the HPA axis, with a corresponding increase in cortisol secretion that will cross the placenta and reach the unborn child [23]. Experimental researches have shown that fetal exposure to elevated glucocorticoid levels has resulted in deleterious effects in the hipocampus [24, 25].

A systematic review examining associations with prenatal cortisol concentrations in humans identified that higher levels of maternal cortisol during pregnancy were linked to low birth weight and shorter length in newborns. Moreover, a higher incidence of respiratory and skin diseases, and lower cognitive and motor development were seen, in addition to psychological and behavioral problems [26].

Studies have also shown that stressed expectant mothers can cause long-term changes in the neurological development of their child. Increased anxiety and decreased attention capacity may be expressed over the course of their development, as was detected in a survey regarding the effects of pre- and post-natal stress on child development, in which high cortisol levels were associated with a larger volume of the right amygdala in girls and the emergence of emotional difficulties [27].

On the other hand, investigations have also shown that OT is elevated in situations of stress, anxiety or depression due to dysregulation of the HPA axis in response to these factors [28]. Within this approach, OT has the function of attenuating the pituitary gland, thus decreasing cortisol secretion. A corroborating study was conducted by Mah et al. [29] within which 24UI of OT, alternating with placebo at a one week interval, was administered to 25 mothers with postpartum depression. The achieved results did not show a positive effect on the depressive symptoms, however, the authors noted a significant improvement in the relationship between mothers and their babies.

This finding reflects the importance of the inclusion of OT in parental care of offspring, increasing their chances of survival. Even if stress and gestational depression interfere negatively in the development of key fetal organs, through cortisol and other catecholamines, OT is responsible in the post-natal period for modulation of these harmful effects on survival and adaptation to the environment. In any event, the evolutionary function of fetal programming is to prepare a child for the specific environment in which it will find itself at birth. Additionally, both evolutionary ecology and molecular biology confirm that a particular genotype can result in different phenotypes under a given environmental condition. Furthermore, the impact of environmental factors experienced by a generation can determine the development and behavior of the next generation (epigenetic inheritance). Therefore, environmental factors during pregnancy can directly affect fetal development, which can last for several generations [30].

The concept of an embryo or fetal adaptive response to an inadequate intrauterine environment, resulting in adverse consequences, is consistent with the definition of "programming" proposed by Lucas in 1991, in which he says "induction, deletion or impaired development of a somatic structure or physiological system through stimulus or insult at a critical period, results in long-term consequences for functional development" [31]. As such, damage that can be caused in the oxytocinergic system during the intrauterine period of embryos and fetuses in mothers enduring stress cannot be discounted, and may be implicated in the emergence of different NCDs in a child's adulthood (Figure1).

In the specific case of a nutritionally poor intrauterine environment, the "metabolic imprinting" it causes in embryos and fetuses is an adaptive response of the organism to adjust to the specific nutritional conditions in the early developmental stages, characterized by: (1) a susceptibility limited to a critical ontogenic window early in development; (2) a persistent effect throughout adulthood; (3) a specific and measurable result, which may be different between individuals; and (4) a dose-dependent or linear relationship, between a specific exposure and outcome [32].

In this sense, it is important to highlight that environmental action on a genotype results in phenotypic plasticity, i.e., it depicts the ability of an individual to adjust their physiology/metabolism and morphology due to the action of environmental factors, generating genetic variability.

This plasticity is essential for biological processes and for the survival of a species, as it can create adaptive advantages in different adverse or unstable environmental conditions, as in the case of fetuses submitted to low nutritional intake. This line of reasoning can be extended to maternal behavior that is deficient in affection and sensory contact with the newborn.

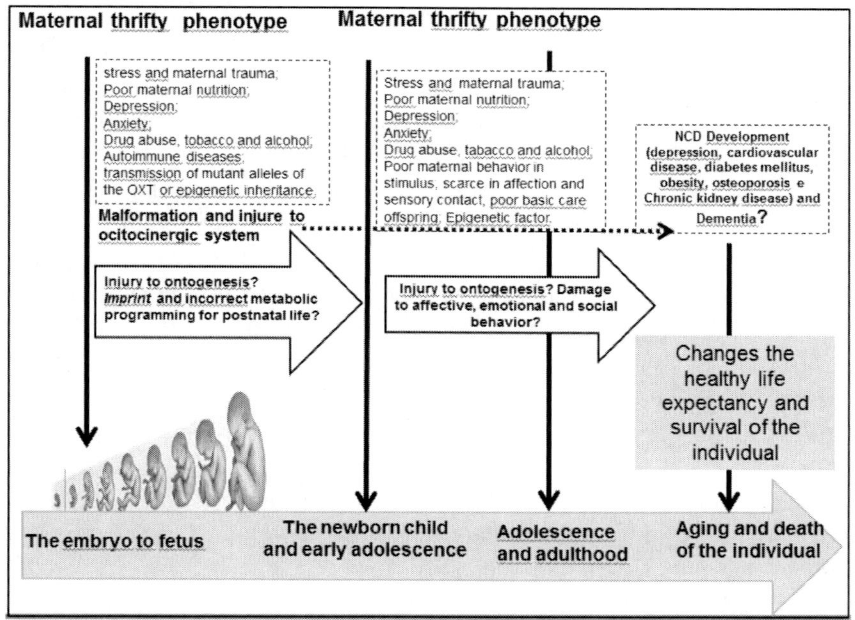

Figure 1. Scheme of hypothetical maternal thrifty phenotype action on ontogenesis and human NCDs.

An adverse environment with little stimulus is likely to be created, which will be of fundamental importance for the organization of behavioral responses during the early stages of child development. The importance of early contact and experience in modulating social behavioral responses over the course of an individual's lifetime has been demonstrated in the literature. Maternal behavior during both the gestational and postnatal periods has also been shown to affect the establishment of an offspring's social relationships and may be caused by a possible alteration in the oxytocinergic system, leading to the development of NCDs.

2. OXYTOCIN RECEPTOR AND OXTR POLYMORPHISM

The central effects of oxytocin are mediated by its receptor, consisting of a protein formed of nine amino acids coupled to the G protein [20]. Additionally, OT receptors are positively regulated by estrogens, and researches indicate that administration of this hormone induces an increase in OT serum levels in a 12-hour period in humans [33]. The gene encoding the oxytocin receptor (OXTR) was first identified in 1992 by Kimura [34] et al., located on the short arm of chromosome 3p25, spanning about 19 kb and containing three exons and 4 introns. OXTR was subsequently completely mapped in humans in 1994 by Inoe et al. [35] OXTR polymorphisms are variations of single nucleotide located in the third intron, and studies suggest they are associated with the emergence of mood and behavioral disorders [36, 37, 38].

Genetic studies indicate that a mutation occurred at some point in human evolution, replacing the basic nitrogen guanine (G) with an adenine (A), as the allele is missing in other primates, leading to the conclusion that the G is the ancestral allele. Data on ethnicity demonstrates that Caucasians are mostly carriers of two copies of the G allele, while Asians present a higher frequency of the A allele and around 40% of these are homozygous for the G allele [39, 40, 41].

Considering that the frequency of alleles differ between races, there will be a difference in phenotypes according to the studied population. For example, the authors of two separate researches, one involving a Chinese children and adolescent population, and the other composed of Japanese individuals, found significant genetic association between the A allele of OXTR polymorphism 2254298 with autism [43, 44]. However, a study conducted in Israel with a Caucasian majority population found a significant association between the G allele and autism [45].

Furthermore, interaction between culture and ethnicity seems to shape the genotype, as shown by Kim [46] et al. when examining the association of cultural norms, psychological disorders and OT receptor gene polymorphism between an American and Korean population. The authors noted that carriers of any homozygous or heterozygous G allele in the American sample had the tendency to seek more emotional support in comparison to the Asian group with the same genotype. No statistically significant differences were found between carriers of the AA genotype in both ethnic groups.

In the absence of early adversity, the AA genotype can be associated with lower levels of depressive symptoms and social anxiety scores, when compared to carriers of the GG genotype, as revealed in studies by Mc Quaid [47] et al. and Tompson [48] et al. A specific genetic variation may predispose the onset of a pathology, although being associated with adversity and environmental contingencies may have beneficial or non-beneficial effects [41, 49] on the behavioral phenotype. The lack of studies associating OXTR polymorphism and gene-environment interaction, however, makes it difficult to draw a conclusion from such information.

Another factor that may contribute to phenotypic differences between genotypes is the association between OXTR polymorphism and amygdala volume. The alteration of this structure can facilitate emergence of depression as it is related to emotional and social memory, involving feelings of friendship, love and affection between people [50].

In a meta-analysis, Hamilton et al. [51] verified through neuroimaging, that non-medicated individuals with major depression presented an abnormally increased amygdala volume.

It is important to highlight that OXTR polymorphism can affect the volume of this limbic structure even at a young age, with a neurobiological hypothesis of the OXTR variant being a risk factor for the development of psychopathology.

Furman et al. [52] analyzed the association between OXTR and amygdala volume and found that individuals homozygous for the G allele had significantly smaller volumes of amygdala in comparison to heterozygous individuals and those with the AA genotype.

These results were similar to those reported by Inoue et al. [53] who reported that individuals homozygous for the A allele showed an increased amygdala size.

Conversely, different results were found in research composed of a sample of Chinese women homozygous for the A allele, who presented with a more pessimistic temperament, although they had a smaller amygdala volume in comparison to individuals with the GG/GA genotype [54].

Based on this information and combined with the embryonic or fetal hypothesis, which postulates that environmental factors (nutrition) have an effect right from the intrauterine stage and program the risk for onset of disease and even premature death in adulthood, it is believed that OXTR polymorphism may also be involved in the etiology of multifactorial diseases, such as depression, Alzheimer, CVD, DM, obesity and osteoporosis.

At the time this hypothesis was proposed by Barker et al., it did not postulate on the involvement of genetic polymorphisms or the lack of affection and maternal well-being influencing the formation and consolidation of the oxytocinergic system, associated with the emergence of NCDs. However, with the advancement of knowledge, this hypothesis can now be enriched with this information, increasing our understanding of the etiology of NCDs. Therefore, this review aims to provide an overview of the association of OT and its genetic polymorphisms with the most prevalent NCDs in the world population, seeking to integrate it to the thrifty phenotype hypothesis throughout the ontogenetic process [55].

3. Oxytocin, OXTR Polymorphism and Chronic Noncommunicable Diseases

3.1. Oxytocin, OXTR Polymorphism and Depression

The prevalence of depression has reached high proportions, turning it into a major public health issue. It especially affects the elderly, women, people with chronic diseases, and those with greater social vulnerability. This disorder is currently the second highest cause of disability in the 15-44 year age group, with warnings from the World Health Organization that by 2020 this disease could move from fourth to second place in terms of major NCDs [56, 57].

Glucocorticoid hormones have a significant part to play in the expression of human behavior. One of the mechanisms related to the pathophysiology of depression may be linked to changes in the ability of circulating glucocorticoids to exert their negative feedback on secretion of HPA axis hormones [58, 59].

Hyperactivity of the HPA axis in major depression is one of the most consistent findings in the pathophysiology of this disease. Patients with depressive disorders present elevated cortisol concentrations in plasma, urine and cerebrospinal fluid, as well as hypertrophy of the adrenal glands and increased pituitary gland volume. It is postulated that the hypothalamus undergoes a dysfunction of its structures during the episode of depression, resulting in activation of the HPA axis, increasing the release of corticotropin (Adrenocorticotropic hormone; ACTH) by the anterior pituitary gland, which in turn stimulates the adrenal cortex to produce cortisol [60]. High concentrations of the cortisol hormone cause decreased libido and appetite, psychomotor alterations and sleep disorders, in addition to anatomical and functional changes of hippocampal neurons with suppression of neurogenesis and neuronal death [61, 62, 63, 64].

Studies have demonstrated that OT can be involved in HPA inhibition in response to stressful events, thus reducing the secretion of cortisol [65, 66] one such research noted that central administration of a low-dose OT in rodents was able to cause attenuation of the HPA axis, induced by psychological stress [67]. Another hypothesis that implies OT action on the pathophysiology of depression is the supposition that OT has the property to induce an increase in serotonergic activity, inhibiting antagonists of the serotonin 2A/C receptor and exerting anxiolytic effects. This was demonstrated in an animal model research through the central application of OT injections in mice with serotonin receptor deficiency, facilitating the release of this hormone [68].

OT has a physiological regulatory function on the HPA system and may, therefore, present in reduced concentrations in individuals with major depression. This can have an effect on some disease symptoms, such as social isolation, appetite loss, decreased libido and sexual desire. In this context, Scantamburlo [69] et al. found a negative correlation in measures of OT plasma concentrations in patients diagnosed with major depression (MD), that is, individuals with a higher Hamilton Rating Scale score showed lower measurement values of oxytocin, similar to results described by Ozoy [70] et al. in patients diagnosed with MD.

In some cases, however, the concentration of OT is the opposite, namely, the OT appears in high concentrations in individuals with depression. This proved to be the case in research by Parker [65] et al., in which depressed subjects had significantly elevated OT serum levels in comparison to the healthy control group, corroborating the results of Purba71 et al., where an increased expression of oxytocin mRNA was found in individuals with depression, as compared to the controls. These findings reinforce the role of OT as a biomarker of social isolation in major depression.

A genetic variation in the oxytocin receptor gene (OXTR) has been associated with phenotypes related to mood disorders, mainly depression, suggesting that people carrying the A allele of OXTR polymorphism are more susceptible to developing these disorders [41].

Researchers have concluded that people who are homozygous for the G allele are more optimistic, have higher self-esteem and present greater empathy, while those with the A allele are more susceptible to the emergence of psychiatric disorders [72]. Rodrigues et al. [73] verified in a population of 192 students from the University of California, that carriers of the GG genotype presented more empathy and a greater ability to detect emotion through facial expression, in addition to having a prosocial temperament. Luch [74] et al. tested the hypothesis of associations between positive and negative affects, and social and emotional loneliness with OXTR in healthy subjects, carriers of the AA genotype, and noted a greater negative affect.

On the other hand, other studies [75, 76] have found contrary results, like Mc Quaid76 et al. who observed that carriers with two copies of the G allele, who suffered maltreatment in childhood, presented greater symptoms of depressive disorders, underlining the belief that OXTR polymorphisms may suffer environmental modulation. This suggests that individuals who carry genetic variations associated with stressful events are more vulnerable to psychological disorders [77]. At the same time, they may also be stronger if exposed to a stable social environment, as seen in research with carriers of the G allele who were raised in a stable family environment and showed greater resilience and a positive affect [78].

3.2. Oxytocin, OXTR Polymorphism and Cognition

Alzheimer's disease is one of the major causes of dementia in the world. The prevalence of this disease in developed countries is approximately 1.5% at 65 years of age, reaching on average around 30% by 80 years. In countries where the number of elderly has been rising, however, the frequency of Alzheimer's has increased steadily [5]. Aging of the brain is associated with decreased neuronal groups of the cortex and subcortex areas, thus contributing to symptoms of dementia. Genetic studies have also demonstrated the influence of genetic polymorphisms on increased risk of developing Alzheimer's, especially Apolipoprotein E (APOE) polymorphism [79].

OT receptors are found in several regions of the brain, including the hippocampus, a region associated with the process of learning and memory because of its close relationship with the HPA axis. In this scenario, stress is known as a critical regulator of brain and cognition functions, since cortisol has negative effects on the hippocampus, such as neurogenesis suppression, increasing cognitive decline as a consequence [80].

Within this context, OT has been implicated in the social cognitive processes by virtue of its ability to promote the perception of social-emotional information. Studies have shown that the administration of intranasal OT has positive effects on social memory, facilitating the processing of social stimuli [81].

Oxytocin plays a significant role in recognition of the same species, a vital component of cognitive development in humans and other animals that live in social groups [82]. Research linked to the hypothesis of an association between OXTR polymorphism and facial memory, involving two sample groups from households in the United Kingdom and Finland, demonstrated that in both samples, approximately one third of individuals who were homozygous for the A allele were associated with a poorer memory for facial recognition [83].

Only one study investigating the association between OT and dementia syndrome, and involving the analysis of human post-mortem brain tissue was found. The sample was composed of 23 subjects with Alzheimer's disease and six control subjects, with measurement made of OT mRNA expression. The authors found no association between Alzheimer's disease and mRNA expression in this sample [84]. Above all it is important to highlight the scarcity of research involving OT, OXTR polymorphism, and cognitive function or Alzheimer's in humans, making a reasonable summary of this theme difficult.

3.3. Oxytocin, OXTR Polymorphism and Obesity

The World Health Organization has now come to consider obesity as a public health problem that is just as concerning as malnutrition, regardless of the economic situation of the countries involved, as both developed and developing nations have experienced this challenge. Obesity has several interrelated factors, such as metabolism, sedentary lifestyle and poor eating habits, together with genetic characteristics and their interaction with the environment (culture, education, interpersonal relationships) [85].

Research is currently focused on the role of the oxytocin gene in energy metabolism. A study involving rodents with an OT receptor deficiency showed that they presented larger abdominal fat deposits [86]. One of the principle hypotheses for the association of OT with basal metabolism relates to increased lipolysis and thermogenesis, as well as the action on insulin. Such findings were demonstrated in a research in which intracerebral OT infusion took place in obese rodents. By the end of the study, the animals presented decreased body weight, increased adipose tissue lipolysis and fatty acid β-oxidation and reduced glucose intolerance and insulin resistance [87, 88].

Another mechanism of OT action on obesity is the association with oleoylethanolamide (OEA), an ethanolamine fatty acid formed from oleic acid and phosphatidylethanolamine in the brain and intestine, which performs functions affecting appetite, inducing satiety and reducing body weight gain. An experimental study in which Sprague-Dawley rats were fed a diet rich in saturated fat (group 1), extra virgin olive oil (group 2), and a low-fat diet (group 3) found that mice consuming an olive oil based diet showed an increase of OEA in plasma and a greater expression of OT mRNA in the hypothalamus [89]. Romano [90] et al. suggested that oleoylethanolamide could have a stimulating effect on OT neurosecretion. Deblon [87] et al. also reinforce this concept, stating that it is not only OEA infusion that can modulate OT levels, but a microinfusion of this hormone can also increase the production of this fatty acid.

A study conducted with dyslipidemic rodents treated with subcutaneous OT for two weeks observed a decreased diameter of adipocytes and increased epididymal adipose tissue protein content, although no differences in weight were noted [91].

A further report linking OT with body composition suggests that oxytocinergic receptors in the hypothalamus participate in food intake regulation, exerting anorexigenic effects. These effects take place in the postprandial period, given that their levels reach high peaks after a meal, decreasing peak glucose response to food intake, and contributing to the control of reward-related eating [92].

Ott [92] et al. observed that administration of intranasal OT in a population of twenty healthy men decreased the consumption of food with sweet flavors, such as chocolate cookies. Furthermore, OT promoted induced satiety by modulating gastric distention, acting together with the GABAergic synapses. It is also speculated that the presence of OT receptors in the ventral tegmental area of the brain may interfere with dopamine signaling in the nucleus accumbens, which contributes to regulation of food intake of pleasantly palatable foods, such as carbohydrates and sweet foods [93].

Moreover, there is an association between OT and leptin, a hormone known to stimulate satiety and regulate thermogenesis, as noted in animal model researches, where OT stimulated the production of leptin in rodents deficient of this hormone [94].

Authors of further research involving the administration of leptin in rats showed a significant increase of OT mRNA expression by the end of the experiment [95]. Kujath [96] et al. also verified a positive association between serum levels of oxytocin and leptin in a group of diabetic women, together with a reduction in waist circumference. However, until the present time, no studies focused on the association between OXTR polymorphisms and obesity have been found in the literature.

3.4. Oxytocin and OXTR Polymorphism in Modulation of Carbohydrate Metabolism: Focus on Diabetes Mellitus

Diabetes Mellitus is considered aNCD with great relevance and impact on public health and society due to its consequences, generated by functional disability and early retirement. Type 2 diabetes comprises 90% of the DM present in the world and is closely linked to overweight and a sedentary lifestyle [97]. According to World Health Organization estimates, more than

180 million people have been diagnosed with DM and this number will probably rise by 2030, with the increase occurring principally among older age groups [98]. Environmental and genetic circumstances give rise to some of the risk factors for onset of this disease, which is characterized by metabolic stress due to insulin deficiency and consequent hyperglycemia, with serious complications [99].

In this context, because OT receptors are also present in the pancreas, it is speculated they have a role in regulation of blood glucose, promoting glucose oxidation. Oxytocin receptors are expressed in the islets of Langerhans where they stimulate the secretion of insulin and glucagon. Human studies have shown that diabetics present a lower concentration of OT, which corroborates this hypothesis [100, 101].

Zhang [102] et al. observed that treatment with OT significantly improved fasting glucose intolerance and insulin levels in the blood of mice; another study with diabetic rodents verified a decrease in the OT receptor gene [103]. OT promotes insulin secretion through activation of cholinergic neurons that innervate pancreatic β cells, stimulate phosphoinositide metabolism, and consequently, an activation of protein kinase C, which are responsible for catalyzing dephosphorylation reactions [104].

Another factor that can contribute to blood glucose reduction is the anorexigenic effect of OT for foods high in sugar, as was demonstrated by an experimental research administering oxytocin to monkeys. The authors observed a reduction in intake of foods rich in sugars, especially those rich in fructose [105].

In a study by Herisson et al., a blood–brain barrier penetrant OT receptor antagonist (L-368,899) was administered in mice and a high-carbohydrate diet was offered soon afterwards. The results showed an increase in consumption of these foods, demonstrating that oxytocin inhibits the appetite for carbohydrates. Furthermore, it was found that sucrose consumption considerably increased the expression of OT gene receptors, suggesting a functional relationship between OT and sugar intake [106].

On the other hand, studies have shown that intraventricular infusion of OT receptor agonists caused a dose-dependent decrease of food consumption in rats. Amico [107] et al. noted that deletion of the OT gene in mice is associated with high intake of sucrose solutions, results that are similar to those found by Sclafani [108] et al., where the absence of OT resulted in increased daily intake of sweets and carbohydrates in rodents [109].

Another possible explanation for the role of OT in diabetes could be its close relationship with dopamine, with eating behavior being linked to the sensation of pleasure and reward [110] *In vitro* research has shown that dopamine D1 (DRD1) and D2 (DRD2) receptor agonists can facilitate the central and peripheral secretion of OT, i.e., the stimulation of receptors, in particular DRD2, may regulate OT secretion [111] In this regard, a research demonstrated that individuals with the A1 allele, the polymorphism of the D2 dopamine receptor gene, are associated with diabetes mellitus. People with the dopamine gene A1 allele tended to consume a larger number of carbohydrate and sweets [112].

The literature is still scarce on this issue with respect to OT and DM, and few studies have reported an association between the two. Additionally, no study was found relating to an association between OXTR polymorphism and DM.

3.5. Oxytocin, OXTR Polymorphism and the Cardiovascular System

Diseases of the circulatory system are the leading causes of death worldwide, according to the World Health Organization, accounting in 2009 for 28.7% of deaths in developing countries and 26.6% in developed countries [113]. Oxytocin when released into the CNS affects multiple areas involved in the control of the cardiovascular system, and peripherally secreted oxytocinergic receptors are associated with oxytocinergic neuronal hemodynamic modulation.

OT receptors are also expressed in aortic artery cells and cardiac muscle, further adding that OT induces the release of atrial natriuretic peptide (ANP), a hormone synthesized primarily by cardiomyocytes. The function of this hormone is to control blood pressure, diuresis, and to inhibit the release and action of various hormones, including aldosterone, angiotensin II, renin and vasopressin. By release of the ANP, OT produces an ionotropic and chronotropic negative response, leading to a decreased contraction force and heart rate, highlighting the role of OT on cardiovascular and endothelial homeostasis. Furthermore, the OT receptor are coupled to a Gq/n protein, which induces contraction of the myocardium [115].

The daily peripheral administration of OT over 5 days in animal models has proven to be effective in lowering arterial blood pressure. Thus, women who receive intravenous infusion of OT to reduce bleeding during childbirth present hypotension due to the decrease of vascular resistance [116] Light [117] et al. also demonstrated that frequent hugs between spouses/partners are associated with lower blood pressure and higher OT levels in premenopausal women. These results can be explained by a reduction in central adrenergic activity, mediated by oxytocin, and exerting peripheral effects on the vascular endothelium.

Human clinical trials have demonstrated the function of OT as an anti-stress hormone, modulating the HPA axis by exerting direct effects on sympathetic and parasympathetic preganglionic neurons and attenuating the decrease of sympathetic activity during stress, such as tachycardia and increased blood pressure, as well as the reduction of circulating cortisol [118].

Studies have also shown that OXTR polymorphism seems to be related to responses to stressful events. This is verified in a study where individuals with two copies of the G allele presented a decrease of pre-ejection period, representing a greater sympathetic system response to the presence of stressors. The same subjects, however, had lower cortisol concentrations upon waking [119]. This result is corroborated in research by Chen et al. [120] where subjects who were carriers of the GG genotype had lower cortisol levels in response to stress, suggesting that these individuals are more protected from cardiovascular risk. However, little research has taken place to verify the role of OXTR polymorphism in CVD.

OT also acts on the endothelium, in addition to its known cardiovascular functions. Deblon et al. [87], following intracerebral infusion of OT in rodents, reported an increased mRNA expression of stearoyl-coenzyme A desaturase 1 (SCD1), an enzyme that plays a key role in the cellular biosynthesis of fatty acids. A further study conducted to verify the effect of OT administration on atherosclerosis in rabbits showed a significant reduction in accumulation of fatty plaques, cholesterol and other substances in artery walls through the inhibition of C-reactive protein [121].

Moreover, although without significant difference, after OT administration the adipose tissue of these animals presented a broad expression of adiponectin mRNA, a hormone linked to the control of food intake and energy homeostasis, as well having a protective effect against atherosclerosis [122].

Another association of interest is the action of OT as an inhibiter of oxidative stress, defined by an imbalance between the production of reactive oxygen species (ROS) and endogenous antioxidant capacity. This can have a key role in the etiology of cardiovascular diseases [123]. *In vitro* experiments have demonstrated that OT inhibits nicotinamide adenine dinucleotide phosphate (NADPH) oxidase generation, the main source of ROS production. Once activated, NADPH results in the production of superoxide anion and its excessive production is harmful to the body, being the basis of many NCDs [124].

Szeto et al. [125] evaluated the effect of OT on superoxide production in the vascular wall, mediated by NADPH oxidase activity. A reduction was observed in endothelial cells and aortic and smooth muscles, and a decrease in production of IL-6 in macrophages was also noted. This suggests a potentially greater role of oxytocin in the control of oxidative stress and inflammation, factors that are closely related to the development of atherosclerosis.

3.6. Oxytocin, OXTR Polymorphism, Musculoskeletal Tissue Modulation and Osteoporosis

Osteoporosis is considered an important public health problem, with projections of 6.3 million cases expected by 2050. This disease can have detrimental effects on quality of life, with a significant decrease in mobility and even loss of autonomy. The strength and integrity of the skeletal system depends on the balance between osteoclasts (resorption) and osteoblasts (formation). Bone tissue *in vitro* experiments have demonstrated that these cells express OT receptors [126]. This indicates that OT has an important function in bone homeostasis, contributing to the prevention of osteoporosis. Further *in vitro* experiments have also shown that increased expression of OT receptors occurs during osteogenesis [127, 128].

The regenerative capacity of skeletal muscle decreases with age, yet OT plays an indispensable role in the regeneration of muscle tissue. At the same time, OT plasma levels tend to decrease with age. Research [129] has shown, however, that the administration of OT stimulates an increase in muscle regeneration capacity due to activation of NF-KB enzyme signaling (a protein produced by fibroblasts, osteoblasts and mesenchymal cells that is involved in osteoclast differentiation and activation), and mitogen-activated protein kinase (MAPK) protein transduction of central signals in the regulation of bone mass [130].

Elabd [131] et al. evaluated the expression of oxytocin receptors in skeletal muscle of young and old rodents. A significantly higher proportion of receptors were found in young rats, demonstrating a decrease in OXTR expression with aging and partially explaining the reduction of OT circulating in plasma with age. Thereafter, the authors administered subcutaneous OT injections to the old rodents, rapidly increasing muscle regeneration by activating the MAPK. Rodents deficient in OT receptors presented signs of premature aging of muscle tissue, as well as reduced muscle fibers and mass, increasing the risk of sarcopenia onset. This data suggests that the lack OT causes a decline in the myogenic responses of satellite cells, leading to reduced maintenance of muscle tissue. These same authors administered OT in rats after oophorectomy, reversing the bone loss and leading to a significant improvement in bone microarchitecture and biomechanical strength [132].

Studies [128, 133] involving humans have revealed that OT is strongly associated with osteoporosis, since women with this disease have a lower concentration of this hormone in blood plasma. In addition, no research was found linking OXTR polymorphism and osteoporosis.

3.7. Oxytocin, OXTR Polymorphism and Kidney Disease

Chronic kidney disease (CKD) is a major public health problem with negative outcomes. The main risk factors in particular for the occurrence of CKD are hypertension and diabetes [134, 135]. CKD has received increasing attention from the international scientific community due to its high prevalence, which has been demonstrated in recent studies, as per data from the National Health and Nutrition Examination Survey (NHANES) [4].

This study, conducted between 1999 and 2004 in the United States (USA), involved a sample of over 13,000 adults aged 20+ years. Data analysis revealed that approximately 13% of the US adult population have stage 1 to 4 CKD [4].

OT receptors are also found in the macula densa cells of the kidney, which are involved in osmolarity regulation and producing natriuresis through modulation of the tubuloglomerular feedback. Regulation of sodium occurs in this way, decreasing its reabsorption in the distal tubule, in addition to its known function in regulating natriuretic peptide release [136].

In a series of experiences with experimental models, Gutkowska and Jankowski [136] found that OT controls natriuresis in the kidneys through the activation of nitric oxide synthase (NOS). This in turn leads to the formation of nitric oxide, with a consequent increase in cyclic guanosine monophosphate (cGMP), which is responsible for mediating natriuresis by closing sodium channels.

A clinical trial conducted by Rached et al. with rodents involved the administration of cisplatin in order to induce nephrotoxicity, with a consequent increase in oxidative stress markers. In parallel, the authors applied intravenous doses of OT. At study end, a reduction in the expression of mRNA of NADPH oxidase was noted, in addition to decreased levels of malonyldialdehyde (MDA) and an increase in superoxide dismutase [137].

Similar results were found with Wistar albino rats that underwent renal tissue ischemia. The authors assessed levels of MDA, urea and creatinine in this study and observed increases in the concentrations of these markers. When treated with OT, however, a significant attenuation in these parameters was seen, thus concluding that OT has a large anti-oxidative effect, resulting in improved renal function by reversing nephrotoxicity [138]. Studies with OXTR polymorphism associated with chronic kidney disease have not been previously documented in the literature.

CONCLUSION

Oxytocin is a molecule that plays many essential roles, not only in respect to reproduction, nutrition and emotional and social ties, but also in performing the function of preparing or regulating humans for survival and adaptation to various adverse or non-adverse environments. Specifically, OT seems to be closely involved in human ontogenesis. However, for OT to be synthesized, secreted and able to appropriately exert its broad spectrum of biological functions, it is essential that ideal conditions of development are offered to the embryo and fetus, such as nutrition and maternal well-being.

This is also true for the newborn, as the mother has a fundamental role in the positive regulation of OT through breast milk, affection, and the full range of parental care. Affective and emotional ties, social aggregation, and parental care should remain throughout the life of the individual to give the OT support necessary to maintain physical, mental, emotional and social health. It is therefore believed that OT is a component that suffers a strong influence of the thrifty phenotype, proposed by Hales and Barker. As their receptors are in key organs responsible for maintenance of organic homeostasis, such as the brain, heart, pancreas and kidneys, as well as exerting the role of neurotransmission and neuromodulation in the body. Thus, if the embryo, fetus, newborn and child are kept under adverse conditions, such as nutritional, affective, sensory and social scarcity, it is likely that their full development will be permanently affected.

Malformation of the oxytocinergic system and its inefficiency can jeopardize the development and functioning of key organs, causing the appearance of NCDs in adulthood. However, it is important to highlight that the literature regarding OT and its polymorphism, associated to NCDs, are insufficient and the majority of studies are conducted with experimental models, more directed towards a behavioral approach. Studies involving the OXTR polymorphism related to mood and behavioral disorders are found in the literature, but studies linking this polymorphism with other NCDs, such as dementia, especially Alzheimer's disease, CVDs, diabetes, osteoporosis and chronic kidney diseases are rare. Nonetheless, studies have demonstrated that there is a strong interaction between OXTR polymorphism and environmental factors. In other words, OXTR polymorphism does not act alone as it is influenced by environmental, ethnic, emotional and affective conditions [139].

All these factors can alter OXTR polymorphism expression, affecting the synthesis of oxytocin. Additionally, due to OXTR phenotype plasticity, it can be influenced by stressful conditions, in much the same way as it has an increased likelihood of benefitting from positive experiences.

REFERENCES

[1]   Macinko, J. Dourado, I., Frederico, C., and Guanais, F. C. (2011). Chronic Diseases, Primary Care and Health Systems Performance: Diagnostics, Tools and Interventions New York: Inter-American Development Bank; [on-line]. http://publications.iadb.org

[2]   World Health Organization, Global status report on noncommunicable diseases 2010, Geneva: World Health Organization.

[3]   Capilheira, M. F. Santos, I. S. Azevedo, J. M. R. and Reichert F. F. (2008). Risk factors for chronic non-communicable diseases and the CARMEN Initiative: a population-based study in the South of Brazil. *Cad. Saúde Pública*, 24(12): 2767-2774.

[4]   Centers for Disease Control and Prevention (CDC). (2007). Prevalence of chronic kidney disease and associated risk factors - United States, 1999-2004. *MMWR Morb. Mortal. Wkly*, 256(8):161-5.

[5]   Atalaia Silva, K. C. Ribeiro, P. C. C.,and Lourenço, R. A. (2008). Epidemiologia das demências. *Rev. Hosp. Univ. Pedro Ernesto*, 7(1): 46-1.

[6]   Bliuc, D. Nguyen, N. D., Milch, V. E.,Nguyen, T. V., Eisman,J. A., and Center, J. R. (2009), Mortality Risk Associated With Low-Trauma Osteoporotic Fracture and Subsequent Fracture in Men and Women. *JAMA*, 301(5):513-521.

[7]   Meinster, B. (2007). Neurotransmitters in key neurons of the hypothalamus that regulate feeding behavior and body weight. *Physiol. Behav.*, 10(92):263-71.

[8]   Francis, S. M., Sagar, A., Levin-Decanini, T., Liu, W., Carter, C. S., Jacob S. (2014) Oxytocin and vasopressin systems in genetic syndromes and neurodevelopmental disorders. *Brain Res.*, 14,74-2.

[9]   Kendrick, K. M. (2004). The neurobiology of social bonds. *J. Neuroendocrinol.*, 16(12),1007-8.

[10]   Ross, H. E. and Young, L. J. (2009). Oxytocin and the neural mechanisms regulating social cognition and affiliative behavior. *Front. Neuroendocrinol.*, 30(4):534-47.

[11]   Tost H., Kolachana B., Verchinski B. A. Bilek E. and Goldman AL. (2011) Neurogenetic effects of OXTR rs 2254298 in the exthended limbic system of healthy causasian adults. *Biol. Psychiatry*, 70(9):37-9.

[12]   MacDonald, K. and MacDonald T. M.: The peptide that binds: a systematic review of oxytocin and its prosocial effects in humans. (2010). *Harv. Rev. Psychiatry*,18: 1–21.

[13]   Huffmeijer, R. Van IJzendoorn, M. H. and Bakermans-Kranenburg, M. J. (2013). Ageing and Oxytocin: A Call for Extending Human Oxytocin Research to Ageing Populations – A Mini-Review. *Gerontology,* 59 (1):32-9.

[14]   Clark, C. L. St. John, N. Pasca, A. M. Hyde, S. A. Hornbeak, K. Abramova M. et al. (2013). Neonatal CSF oxytocin levels are associated with parent report of infant soothability and sociability. *Psychoneuroencrinol.*, 38(7):1208-12.

[15]   Todeschin, A. S. Winkelmann-Duarte, E. C. Jacob, M. H. Aranda, B. C. Jacobs S., Fernandes M. C., et al. (2009). 'Effects of neonatal handling on social memory, social interaction, and number of oxytocin and vasopressin neurons in rats. *Horm. Behav.*, 2009 Jun.; 56(1):93-100.

[16]   Zimmerberg, B. and Sageser, K. A. (2011). Comparison of Two Rodent Models of Maternal Separation on Juvenile Social Behavior. *Frontiers in Psychiatry,* 2:39. 17.

[17]   Veenema, A. H. (2012). Toward understanding how earlylife social experiences alter oxytocin and vasopressin regulated social behaviors *Horm. Behav.*, 61(3):304-12.

[18]   Wismer-Fries, A. B. Ziegler, T. E. Kurian, J. R. Jacoris, S. and Pollak, S. D. (2005). Early experience in humans is associated with changes in neuropeptides critical for regulating social behaviour. *Proc. Natl. Acad. Sci. U S A,* 102(47):17237–40.

[19] Alves, A. Fielder, A. Ghabriel, N. Sawyer, M. and Buisman-Pijman F. T. (2015). Early social environment affects the endogenous oxytocin system: a review and future directions. *Front. Endocrinol.*, 11:6-32.

[20] Gimpl, G. and Fahrenholz F. (2001) The oxytocin receptor system: structure, function, and regulation. *Physiol. Rev.*, 81:629–683.

[21] Zhang, G. and Cai, D. (2011). Circadian intervention of obesity development via resting-stage feeding manipulation or oxytocin treatment. *Am. J. Physiol. Endocrinol. Metab.*, 301(5):1004–1012.

[22] Hales, C. N. and Barker, D. J. P. (1992). Type 2 (non-insulin-dependent) diabetes mellitus: the thirfty phenotype hypothesis. *Int. J. Epidemiol.*, 2013; 42(5):1215-22.

[23] Hostinar, C. E. and Megan, R. (2013). Future directions in the study of social relationships as regulators of the HPA axis across development. *J. Clin. Child Adolesc. Psychol.*, 42(4):564-575.

[24] Mulder, E. J. H. Medina, P. G. Huizink, A. C. Van den Bergh B. R. H. and Visser. G. H. A. 2002. Prenatal maternal stress: effects on pregnancy and the (unborn) *Child Early Hum. Dev.*, 70:3-14.

[25] Duthie, L. and Reynolds, R. M. (2013). Changes in the maternal hypothalamic-pituitary adrenal axis in pregnancy and postpartum: influences on maternal andfetal outcomes. *Neuroendocrinology,* 98, 106–115.

[26] Zijlmans, M. A. Riksen-Walraven, J. M. and De Weerth C. (2015) Associations between maternal prenatal cortisol concentrations and ghild outcomes: A systematic review. *Neurosci. Biobehav.*, 53:1-24.

[27] Bus, C. Davis E., P. Shahbaba, B. Pruesner, J. C. Head, K. and Sandman, C. A. (2012). Maternal cortisol over the course of pregnancy and subsequent child amygdala and hippocampus volumes and affective problems. *Proc. Natl. Sci. USA,* (20):1312-9.

[28] Cox, E. Q. Stuebe, A. Pearson, B. Grewen, K. Rubinow, D. and Meltzer-Brody S. (2015). Oxytocin and HPA stress axis reactivity in postpartum women. *Psychoneuroendocrinology,* 55:164-72.

[29] Mah, B. L. Van, I. Jzendoorn, M. H. Smith, R. and Bakermans-Kranenburg M. J. (2013) Oxytocin in postnatally depressed mothers: its influence on mood and expressed emotion. *Progr. Neuro-Psychopharmacol. Biol. Psychiatry,* 40: 267–272.

[30] Liese, A. D. Mayer-Davis, E. J. and Haffner, S. M. (1998) Development of the multiple metabolic syndrome: an epidemiologic perspective. *Epidemiol.*, 20:157-72.

[31] Lucas, A. Programming by early nutrition in man. (1991) *Ciba Found Symp.*, 156:38-50.

[32] Waterland, R. A. Garza, C. (1999). Potential mechanisms of metabolic imprinting that lead to chronic disease. *Am. J. Clin. Nutr.*, 69:179-97.

[33] Nomura, M. McKenna, E. Korach, K. S. Pfaff, D. W. and Ogawa, S. Estrogen receptor- regulates transcript levels for oxytocin and arginine vasopressin in the hypothalamic paraventricular nucleus of male mice. (2002). *Brain Res. Mol. Brain Res.*, 09:84 –94.

[34] Kimura, T. Azuma, C. Saji, F. Takemura, M. Tokugawa, Y. Miki, M., et al. (1992) Oxytocin receptor gene. *J. Steroid Biochem. Mol. Biol.*, 42:253-8.

[35] Inoe, T. Kimura, T. Azuma, C. Inazawa, J. Takemura, M. Kikuchi, T. et al. Structural organization of the human oxytocin receptor. (1994). *J. Biol. Chem.*, 23(51):32451-6.

[36] Zingg, H. H. Laporte, S. A. (2013). The oxytocin receptor. *Trends Endocrinol. Metab.*, 14(5):222-7.

[37] Meyer, L. and Tost, H. (2012). Neural mechanisms of social risk for psychiatric disorders. *Nat. Neurosci.*, 15(5):663-8.

[38] Ebstein, R. P. Israel, S. Chew, S. H. Zhong, S. and Knafo, A. (2010). Genetics of Human Social Behavior. *Neuron*, 65(6):831-44.

[39] Chelala, C. Khan, A. and Lemoine N. R. (2009). SNP nexus: A web database for functional annotation of newly discovered and public domain Single Nucleotide Polymorphisms. *Bioinformatics*, 25(5): 655-66.

[40] Michelini, S. Urbanek, M. Dean, M. and Goldman, D. (1995) Polymorphism and genetic mapping of the human oxytocin receptor gene on chromosome 3. *Am. J. Med. Genet.*, 60(3):183-7.

[41] Brune, M. (2012) Does the oxytocin receptor polymorphism (rs2254298) confer 'vulnerability' for psychopathology or 'differential susceptibility'? Insights from evolution. *BMC Med.*, 10:38.

[42] Wu, S. Jia, M. Ruan, Y. Liu, J. Guo, and Y. Shuang, M. Et al. (2005). Positive association of the oxytocin receptor gene (OXTR) with autism in the Chinese Han population. *Biol. Psychiatry*, 58:74–77.

[43] Liu, X. Kawamura, Y. Shimada, T. Otowa, T. Koishi S., Sugiyama T. Et al. (2010). Association of the oxytocin receptor (OXTR) gene polymorphisms with autism spectrum disorder (ASD) in the Japanese population. *J. Hum. Genet.*, 55:137–141.

[44] Jacob, S. Brune, C. W. and Carter, C. S. (2007). Leventhal B. L., Lord C., Cook E. H. Jr. Association of the oxytocin receptor gene (OXTR) in

Caucasian children and adolescents with autism. *Neurosci. Lett.*, 417: 6–9.

[45] Kim, H. S. Sherman, D. K. Sasaki, J. Y. Xu J., Chu T. Q., Ryu C., Et al. (2010). Culture, distress, and oxytocin receptor polymorphism (OXTR) interact to influence emotional support seeking. *Proc. Natl. Acad. Sci.*, 107:15717–15721.

[46] Thompson, R. J. Parker, K. J. Hallmayer, J. F. Waugh, C. E. and Gotlib I. H. (2011). Oxytocin receptor gene polymorphism (rs2254298) interacts with familial risk for psychopathology to predict symptoms of depression and anxiety in adolescent girls. *Psychoneuroendocrinol.*, 36:144–147.

[47] Mc Quaid, R. J. McInnis, A. O. Stead, J. D. and Matheson K. (2013). A paradoxical association of an oxytocin gene receptor polymorphism: early-life adversity and vulnerability to depression. *Front. Neurosci.*, 2013 Jul. 23;7:128.

[48] Belsky, J. and Beaver, K. M. (2011). Cumulative-genetic plasticity, parenting and adolescent self-regulation. *J. Child Psychol. Psychiat.*, 52:619–626.

[49] Cahill, L. Babinsky, R. Markowitsch, H. J. and McGaugh J. L. (1995). The Amygdala and Emotional Memory. *Nature,* 377(6547):295-6.

[50] Hamilton, J. Siemer. M., and Gotlib, I. (2008). Amygdala volume in major depression disorder: a meta-analysis of magnetic resonance imaging studies. *Mol. Psychiatry*, 13(11):993-100.

[51] Furman, D. J. Chen, M. C. and Gotlib, I. C. H. (2011) Variant in oxytocin receptor gene is associated with amygdala volume. *Psychoneuroendocrinol.*, 36(6):891-7.

[52] Inoue, H. Yamasue, H. Tochigi, M. Abe, O. Liu, X. Takei, K. Et al. (2010). Association between the oxytocin receptor gene and amygdalar volume in healthy adults. *Biol. Psychiatry*, 68(11):1066-72.

[53] Wang, J. Qin, W. Liu, B. Zhou, Y. Wang, D. Jiang, T. and Yu, C. (2014) Neural mechanisms of oxytocin receptor gene mediating anxiety related temperament. *Brain Struct. Funct.*, 219(5):1543-54.

[54] Barker, D. J. Eriksson, J. G. Forsen, T. and Osmond C. (2002) Fetal origins of adult disease: strength of effects and biological basis. *Int. J. Epidemiol.*, 2002, 31:1235-9.

[55] Word Health Association. The ICD-10 Classification of Mental and Behavioural Disorders. (2012) Clinical descriptions and diagnostic guidelines. Geneva: World Health Organization. [On-line serial] http://www.who.int

[56] Brazil. Ministry of Planning, Budget and Management: Brazilian Institute of Geography and Statistics. IBGE. National Health Survey 2013. [On-line serial] http://www.ibge.gov.br/home/estatistica/populacao/pns/2013/default.shtm

[57] Young, A. H. (2006) Antiglucocorticoid treatments for depression. *Aust. N Z J. Psychiatry,* 40: 402-405.

[58] Juruena, M. F. Anthony, T. C. and Pariante, C. M. (2004). The Hypothalamic Pituitary Adrenal axis, Glucocorticoid receptor function and relevance to depression. *Rev. Bras. Pisquiatr.,* 26(3):189-201.

[59] Karim, A. (2013). Brain Physiology and Pathophysiology in Mental Stress. *Physiology,* 2013:1-23.

[60] Hamilton, L. D. and Meston, C. M. (2011). *The role of salivary cortisol and DHEA-S in response to sexual, humorous, and anxiety-inducing stimuli,* 59(5):765-71.

[61] Saraiva, E. M. Fortunato, J. M. S. and Gavina, C. (2005) Oscilações no cortisol na depressão, sono e vigília. *Rev. Port de Psicossomática,* 7(12):1-13.

[62] Hansson, P. B. Murison, R. Lund, A. and Hammar, A. (2015). *Cognitive functioning and cortisol profiles in first episode major depression,* 56(4):379-83.

[63] Talarico, J. N. S. Stress, cortisol levels and coping strategies on memory performance of healthy elderly individuals with mild cognitive impairment and Alzheimers disease [Tese]. University of São Paulo, São Paulo Brazil, 2009.

[64] Parker, K. J. Kenna, H. Á., Zeitzer, J. M. Keller, J. Blasly, C. M. Amico, J. et al. (2010). A. Preliminary evidence that plasma oxytocin levels are elevated in major depression. *Psychiatry Res.,* 178(2):359-62.

[65] Meyer, S. E. Chrousos, G. P. and Gold P. W. (2001) Major depression and the stress system: a life span perspective. *Development and Psychopathology,* 13:565–580.

[66] Windle, R. J. Kershaw, Y. M. Shanks, N. Wood, S. A. Lightman, S. L. and Ingram C. D. (2004). Oxytocin attenuates stress-induced c-fos mRNA expression in specific forebrain regions associated with modulation of hypothalamo-pituitary-adrenal activity. *Journal of Neuroscience,* 24: 2974–2982.

[67] Yochida, M. Takayanagi, Y. Inoe, K. Kimura, T. Yong, L. J. and Onaka T. (2009) Evidence that oxytocin exerts anxiolytic effects via oxytocin receptor expressed in serotonergic neurons in mice. *J. Neurosc.,* 18; 29(7):2259-71.

[68] Scantamburlo, G. Hansenne, M. Fuchs, S. Pitchet, W. Marechal, P. and Pequeux C. (2007). Plasma oxytocin levels and anxiety in patients with major depression. *Psyconeuroendocrino.*, 32(4):407-10.

[69] Ozsoy, S. Esel, E. and Kula, M. Serum oxytocin levels in patients with depression and the effects of gender and antidepressant treatment. (2009). *Psychiatry*, 169(3):249-52.

[70] Purba, J. S. Hoogendijk, W. J. Hofman, M. A. and Swaab, D. F. (1996). Increased number of vasopressin and oxytocin expressing neurons in the paraventricular nucleus of the hypothalamus in depression. *Archives of General Psychiatry*, 53: 137-143.

[71] Saphire, B. S. Way, B. M. Kim, H. S. Sherman, D. K. and Taylor, S. E. (2011) Oxytocin receptor gene (OXTR) is related to psychological resources. *Proc. Natl. Acad. Sci. USA*, 108(37):15118-22.

[72] Rodrigues, S. M. Saslow, L. R. Garcia, N. John, O. P. and Keltner, D. (2009) Oxytocin receptor genetic variation relates to empathy and stress reactivity in humans. *Proc. Natl. Acad. Sci. USA,* 106(50):21437-41.

[73] Lucht, M. J. Barnow, S. Sonnenfeld, C. Rosenberger, A. Grabe, H. J. Schroeder W., et al. (2009). Associations between the oxytocin receptor gene (OXTR) and affect, loneliness and intelligence in normal subjects. *Prog. Neuropsychopharmacol. Biol. Psychiatry*, 33:860–866.

[74] Costa, B. Pini, S. Gabelloni, P. Abelli, M. Lari, L. Cardini A. et al. (2009). Oxytocin receptor polymorphisms and adult attachment style in patients with depression. *Psychoneuroendocrinol.*, 34:1506–1514.

[75] Mc Quaid, R. J. Mclnnis, A. O., Stead, J. D. and Matheson K. (2013).A paradoxical association of an oxytocin receptor gene polymorphism: early-life adversity and vulnerability to depression. *Front. Neurosci.,* 7(128):1-7.

[76] Belsky, J. Jonassaint, C. Pluess, M. Stanton, M. Brummett, B. and Williams R. (2009). Vulnerability genes or plasticity genes? *Molecular Psychiatry,* 14(8):746–754.

[77] Bradley, B. Davis, T. A. Wingo, A. P. Mercer, K. B. and Ressler, K. J. (2013). Family environment and adult resilience: contributions of positive parenting and the oxytocin receptor gene. *Eur. J. Psychotraumatol.*, 18;4:1-29.

[78] Aprahamian, I. Martinelli, J. E. and Yassuda, M. S. (2009) Doença de Alzheimer: revisão da epidemiologia e diagnóstico. *Rev. Bras. Clin. Med.*, 7:27-35.

[79] Dayi, A. Cetin, F. and Sisman, A. R. (2015)The Effects of Oxytocin on Cognitive Defect Caused by Chronic Restraint Stress Applied to

Adolescent Rats and on Hippocampal VEGF and BDNF Levels. Med. Sci. Monit. : *International Medical Journal of Experimental and Clinical Research,* 21:69-75.

[80] Weigand, A. Felser, M. Ejartner, M. Brandt, E. Fan, Y. and Fuje, P. (2013). Effects of intranasal oxytocin prior to encoding and retrieval on recognition memory. *Psychofharmacol.,* 227(2):321-9.

[81] Young, L. J. and Wang, Z. (2004) The neurobiology of pair bonding. *Nat. Neurosci.,* 7(10):1048–1054.

[82] Skuse, D. H. Lori. A. Cubells, J. F. Lee, I. Conneely, K. N. Puura, K. and Young, L. J. (2014). Common polymorphism in the oxytocin receptor gene (OXTR) is associated with human social recognition skills. *Proceedings of the National Academy of Sciences of the United States of America,* 111(5), 1987–1992.

[83] G, Meynen. Unmehopa, M. A. Hofman, D. F. Swaab,W. J. and Hoogendijk, G. (2009). Hypothalamic vasopressin and oxytocin. mRNA expression in relation to depressive state in Alzheimer'sdisease: a difference with major depressive disorder. *J. Neuroendocrinol.,* 21(8):722-9.

[84] World Health Organization. (2015). Obesity and overweight. [On-line serial] http://www.who.int

[85] Takayinagi, Y. Kasahara, Y. Onaka, T. Takahasshi, N. kawada, T. and Nishimori, K. (2008). Oxytocin receptor-deficient mice developed late-onset obesity. *Neuroreport,* 19(9):951-5.

[86] Deblon, N. Veyrat-Durebex C., and Bourgoin, L. (2011). Mechanisms of the Anti-Obesity Effects of Oxytocin in Diet-Induced Obese Rats. Lobaccaro J-MA, ed. *PLoS ONE,* 2011; 6(9):e2556588.

[87] Zhang, G. Bai, H. Zhang, H. Dean, C. and Wu, Q. (2011). Neuropeptide exocytosis involving synaptotagmin-4 and oxytocin in hypothalamic programming of body weight and energy balance. *Neuron,* 69:523–535.

[88] Sospedra, I. Moral, R. Escrich, R. Solanas, M. Vela, E. and Escrich, E. (2015) Effect of High Fat Diets on Body Mass, Oleylethanolamide Plasma Levels and Oxytocin Expression in Growing Rats. *J. Food Sci.,* 80(6):H1425-31.

[89] Romano, A. Cassano, T. Tempesta, B. Cianci, S. Dipasquale, P. Coccurello. (2013). The satiety signal oleoylethanolamide stimulates oxytocin neurosecretion from rat hypothalamic neurons. *Peptides,* 16:21–6.

[90] Eckertova, M. Ondrejcakova, M. Krsova, K. Zorad, K. and Jezova, D. (2011) Subchronic treatment of rats with oxytocin results in improved

adipocyte differentiation and increased gene expression of factors involved in adipogenesis, *Br. J. Pharmacol.*, 162(2):452-63.

[91] Ott, V. Finlayson, G. Lehnert, H. Heitmann, B. Heinrichs, M. Born J. Et al. (2013). Oxytocin Reduces Reward-Driven Food Intake in Humans. *Diabetes,* 62(10):3418–3425.

[92] Berridge K. C. (2009). 'Liking' and 'wanting' food rewards: brain substrates and roles in eating disorders. *Physiol. Behav.*, 97:537–550.

[93] Blevins, J. E. and Baskin, D. G. (2015) Translational and therapeutic potential of oxytocin as an anti-obesity strategy: Insights from rodents, nonhuman primates and humans. *Br. J. Pharmacol.*, [Epub ahead of print].

[94] Perello, M. and Raingo, J. (2013). Leptin Activates Oxytocin Neurons of the Hypothalamic Paraventricular Nucleus in Both Control and Diet-Induced Obese Rodents, ed. *PLoS ONE*, 2013; 8(3):e59625.

[95] Kujath, A. S. Quinn, L. and Elliott, M. E. Varady, V. A. LeCaire T. J., Carter C. S., et al. (2015). Oxytocin levels are lower in premenopausal women with type 1 diabetes mellitus compared with matched controls. *Diabetes Metab. Res. Rev.*, 31(1):102-12.

[96] Schmidt, M. I. Duncan, B. B. E., Silva, G. A. Menezes, A. M. Monteiro, C. A. and Barreto, S. M. (2011). Chronic noncommunicable diseases in Brazil: burden and current challenges. *Lancet,* 377(9781):1949-61.

[97] World Health Organization. Diabetes. [On-line serial] http://www.who.int/mediacentre

[98] Maejima, Y. Iwasaki, Y. Yamahara, Y. Kodaira, M. Sedbazar, U. and Yada, T. (2011). Peripheral oxytocin treatment ameliorates obesity by reducing food intake and visceral fat mass. *Aging (Albany NY),* 3(12):1169-1177.

[99] Elabd, S. K. Sabry, I. Mohasseb, M. and Algendy, A. (2014). Endocr Regul. *Oxytocin as a novel therapeutic option for type I diabetes and diabetic osteopathy*, 48(2):87-102.

[100] Plante, E. Menaouar, A. Danalache, B. A. Yip, D. Broderick, T. L. Chiasson J. L. et al. (2015). Oxytocin Treatment Prevents the Cardiomyopathy Observed in Obese Diabetic Male db/db Mice. *Endocrinol.*, 156(4):1416-28.

[101] Zhang, H. Wu, C. Chen, Q. Chen, X. Xu, Z. Wu, J. et al. (2013). Treatment of Obesity and Diabetes Using Oxytocin or Analogs in Patients and Mouse Models. *PLoS ONE*, 8(5): e61477.

[102] Gutkowska, J. Broderick, T. L. Bogdan, D. Wang, D. Lavoie, J.-M. Jankowski, M. (2009). Downregulation of oxytocin and natriuretic

peptides in diabetes: possible implications in cardiomyopathy. *The Journal of Physiology*, 587:4725-4736.

[103] Gao, Z. Y. Drews, G. and Henquin, J. C. (1991). Mechanisms of the stimulation of insulin release by oxytocin in normal mouse islets. *Biochem. J.*, 276:169–174.

[104] Blevins, J. E. Graham, J. C. Morton, G. J. Bales, K. L. Schwartz, M. W. and Baskin, D. G. (2015). Chronic oxytocin administration inhibits food intake, increases energy expenditure, and produces weight loss in fructose-fed obese rhesus monkeys. *Am. J. Physiol. Regul. Integr. Comp. Physiol.*, 1;308(5):R431-8.

[105] Herisson, F. M. Brooks, L. L. Waas, J. R. Levine, A. S. and Olszewski, P. K. (2014) Functional relationship between oxytocin and apetite for carbohydrates versus saccharin. *Neuroreport.*, 25(12):909-14.

[106] Amico, J. A. Vollmer, R. R. Cai, H. M. Miedlar, J. A. and Rinaman L. (2005). Enhanced initial and sustained intake of sucrose solution in mice with an oxytocin gene deletion. *Am. J. Physiol. Regul. Integr. Comp. Physiol.*, 289:R1798-R1806.

[107] Sclafani, A. Rinaman, L. Vollmer, R. R. and Amico, J. A. (2007). Oxytocin knockout mice demonstrate enhanced intake of sweet and nonsweet carbohydrate solutions. *Am. J. Physiol. Regul. Integr. Comp. Physiol.*, 292(5):1828-33.

[108] Olszewski, P. K. Klockars, A. Olszewska, A. M. Fredriksson, R. Schiöth, H. B. and Levine, A. S. (2010). Molecular, Immunohistochemical, and Pharmacological Evidence of Oxytocin's Role as Inhibitor of Carbohydrate But Not Fat Intake. *Endocrinology*, 51(10):4736-4744.

[109] Winkler, J. K. Woehning, A. Schultz, J. H. Brune, M. Beaton, N. Challa, et al. (2012). TaqIA polymorphism in dopamine D2 receptor gene complicates weight maintenance in younger obese patients. Nutrition, 28(10):996-1001.

[110] Amico, J. A. Layden, L. M. Pomerantz, S. M. (1993). Cameron J. L. Oxytocin and vasopressin secretion in monkeys administered apomorphine and a D2 receptor agonist. *Life Sci.*, 52(15):1301-9.

[111] Barnard, N. D. Noble, E. P. (2009). Ritchie T., Cohen J., Jenkins D. J. A., Turner-McGrievy D2 Dopamine receptor Taq1A polymorphism, body weight, and dietary intake in type 2 diabetes, *Nutrition*, 25(1), 58–65.

[112] World Health Organization (WHO). Noncommunicable Diseases Country Profiles. [On-line serial] http://www.who.int

[113] Costa, F. S. Cunha, S. P. Reis, F. J. Rodrigues, J. A. and Rocha, R. S. (2007) Nitric oxide and atrial natriuretic peptide in the prediction of pregnancy complications. *Rev. Bras. Ginecol. Obstet.,* 29(1): 41-47.

[114] Japundžić-Žigon, N. (2013). Vasopressin and Oxytocin in Control of the Cardiovascular System. *Current Neuropharmacology,* 11(2):218-230.

[115] J. B. Sartain, J. J. Barry, P. W. Howat, D. I. and McCormack. M., Bryant Intravenous oxytocin bolus of 2 units is superior to 5 units during elective Caesarean section *British Journal of Anesthesia,* 101 (6) (2008), pp. 822–826.

[116] Light, K. C. Grewen, K. M. Amico, J. A. (2005). More frequent partner hugs and higher oxytocin levels are linked to lower blood pressure and heart rate in premenopausal women. *Biol. Psychol.,* 69(1):5-21.

[117] Norman, G. J. Cacioppo, J. T. Morris, J. S. Malarkey, W. B. Berntson, G. G. and Devries A. C. (2011). Oxytocin increases autonomic cardiac control: moderation by loneliness. *Biol. Psychol.,* 86(3):174-80.

[118] Norman, G. J. Hawkley, L. Luhmann, M. Ball, A. B. Cole, S. W. and Cacioppo J. T. (2012). Variation in the oxytocin receptor gene influences neurocardiac reactivity to social stress and HPA function: a population based study, 61(1):134-9.

[119] Chen, F. S, Kumsta, R. Von Dawans, B. Monakhov, M. Ebstein, R. P. and Heinrichs, M. (2011). Common oxytocin receptor gene (OXTR) polymorphism and social support interact to reduce stress in humans. *Proc. Natl. Acad. of Sci. USA,* 108(50):19937-19942.

[120] Szeto, A. Rossetti, M. A. Mendez, A. J. Noller, C. M. Herderick, E. E. Gonzales, J. A. et al. (2013). Oxytocin Administration Attenuates Atherosclerosis and Inflammation in Watanabe Heritable Hyperlipidemic Rabbits. *Psychoneuroendocrinology,* 38(5):685–693.

[121] Hoyda, T. D. Fry, M. Ahima, R. S. and Ferguson, A. V. (2007). Adiponectin selectively inhibits oxytocin neurons of the paraventricular nucleus of the hypothalamus. *J. Physiol.,* 585(:805-816.

[122] Reis, J. S. Veloso, C. A. Mattos, and R. T. Purish, S. Machado-Nogueira, J. A. (2008). Oxidative stress: a review on metabolic signaling. *Arq. Bras. Endocrinol. Metab.,* 52(7):1096-1105.

[123] Rabelo, L. A. Souza, V. N. Fonseca, L. J. S. and Sampaio, W. O. Redox. (2010). Unbalance: NADPH Oxidase as Therapeutic Target in Blood Pressure Control. *Arq. Bras. Cardiol.,* 2010; 94(5):643-651.

[124] Szeto, A. Nation, D. A. Mendez, A. J. (2008). Dominguez-Bendala, J., Brooks, L. G., Schneiderman, N., McCabe, P. M. Oxytocin attenuates NADPH-dependent superoxide activity and IL-6 secretion in

macrophages and vascular cells. *Am. J. Physiol. Metab.*, 295(6):E1495-E1501.

[125] Zallone A. (2002). Human osteoclasts express oxytocin receptor. *Biochem. Biophys. Res. Commun.*, 297:442–445.

[126] Breuil V., Panaia-Ferrari P., Fontas E., Roux C., Kolta S., Eastell R., et al. (2014). Oxytocin, a new determinant of bone mineral density in post-menopausal women: analysis of the OPUS cohort. *J. Clin. Endocrinol. Metab.*, 99(4):E634–41.

[127] Beranger, G. E. Djedaini, M. Battaglia, S. Roux, C. H. Scheideler, M. Heymann, D. et al. (2015). Oxytocin Reverses Osteoporosis in a Sex-Dependent Manner. *Frontiers in Endocrinology*, 6, 81.

[128] Elabd, C. Cousin, W. Upadhyayula, P. Chen, R. Y. Chooljian, M. S. Li, J., et al. (2014). Oxytocin is an age-specific circulating hormone that is necessary for muscle maintenance and regeneration. *Nat. Commun.*, 5:4082.10.

[129] Aleksic, V. Aoki, A. Iwasaki, Takasaki A. A., Wang C. Y., Asiko Y. et al. (2010). Low level Er. Yag Laser irradiation enhances osteoblast proliferation through activation of MAPK/ERK Lasers. *Med. Sci.*, 25(4):559-69.

[130] Elabd, C. Basillais, A. Beaupied, H. Breuil, V. Wagner, N. Scheideler, M. Et al. (2008). Oxytocin controls differentiation of human mesenchymal stem cells and reverses osteoporosis. *Stem Cells*, 2008; 26:2399–2407.

[131] Breuil, V. Amri, E. Z. Panaia-Ferrari P., Testa, J. Elabd, and C. Albert-Sabonnadière, C. (2011). Oxytocin and bone remodelling: relationships with neuropituitary hormones, bone status and body composition. *Joint Bone Spine,* 78(6):611-5.

[132] Guillermo, G. G. and, Vivekanand, J. H. A. Doença renal em populações desfavorecidas. DRC em Populações Desfavorecidas. (2015). *J. Bras. Nefrol.*, 37(1):14-18.

[133] Bastos, R. M. R. Bastos, M. G. and Teixeira, M. T. B. (2007). Chronic kidney disease and the challenge to primary health care for early detection. *Rev. APS*, 10(1): 46-55.

[134] Bodi. V., Sanchis. J., Nunez. J., Mainar, L. Minana, G. Benet, I. et al. Uncontrolled immune response in acute myocardial infarction: unraveling the thread. (2008) *Am. Heart J.*, 156:1065–1073.

[135] Gutkowska, J. Jankowski, M. and Antunes-Rodrigues, J. (2014). The role of oxytocin in cardiovascular regulation. *Braz. J. Med. Biological Res.*, 47(3):206–214.

[136] Rashed, L. A. Hashem, R. M. and Soliman, H. M. (2011). Oxytocin inhibits NADPH oxidase in cisplatin-induced nephrotoxicity. *Biomed. Pharmacother.*, 65(7):474-80.

[137] Tuğtepe, H., Sener G. Biyikli, N. K. Yüksel, M. Cetinel, S. Gedik, N. Et al. (2007). The protective effect of oxytocin on renal ischemia/ reperfusion injury in rats. *Regul. Pept.*, 140(3):101-8.

[138] Jacondino, C. B. Borges, and C. A. Gottlieb, M. G. V. (2014) Association of oxytocin receptor gene polymorphisms rs53576 and rs2254298 with depression: a systematic review, *Sci. Med.*; 24(4):411-419.

In: Advances in Oxytocin Research
Editor: Evan Anderson

ISBN: 978-1-63483-991-4
© 2015 Nova Science Publishers, Inc.

Chapter 2

OXYTOCIN IN CARDIOVASCULAR CONTROL: AN UPDATE

Maja Lozić[1], David Murphy[2] and Nina Japundžić-Žigon[1]
[1]Institute of Pharmacology, Clinical Pharmacology and Toxicology,
School of Medicine, University of Belgrade, Belgrade, Serbia
[2]Molecular Neuroendocrinology Research Group,
The Henry Wellcome Laboratories for
Integrative Neuroscience and Endocrinology,
University of Bristol, Bristol, UK

ABSTRACT

Experimental evidence accumulated over the last few decades indicates that the nonapeptide hormone oxytocin, which is synthesized both centrally in the hypothalamus, and peripherally in the heart and the vasculature, can play a beneficial role in cardiovascular control. In the periphery, oxytocin has been found to be a growth and differentiation factor in the developing heart that can also promote regeneration and cardiac healing in ischemia. Both oxytocin released in to the blood and oxytocin derived from the heart can increase the release of atrial natriuretic peptide and hence affect hydromineral homeostasis. Quiescence of oxytocin neurons in the hypothalamic paraventicular nucleus, and the concomitant hyperactivity of vasopressin neurons in the paraventicular nucleus, has been documented to underlie osmotic adjustment to pregnancy. In most vascular beds, oxytocin was found to exert weak vasoconstriction by the stimulation of vasopressin V1a receptors. In contrast, the vasoconstriction of the umbilical artery at term by oxytocin is strong. In the vascular beds of the lungs and the brain, oxytocin can produce vasodilatation via NO production. In the central nervous system, oxytocin acts in a paracrine and autocrine fashion, as well as a classical neurotransmitter impinging either directly in the brainstem on neurogenic cardio-respiratory control, or indirectly by the modulation of behavior and emotions. In the paraventicular nucleus, oxytocin receptors were reported to increase baro-receptor reflex sensitivity, reduce blood pressure short-term variability and buffer the cardiovascular response to stress. Thus, recent data suggests multiple roles for oxytocin in cardiovascular regulation. Further research may reveal therapeutic potential.

Keywords: oxytocin, blood pressure, heart rate, baro-receptor reflex, stress, pregnancy

INTRODUCTION

Nomen (Non) Est Omen

When Sir Henry Dale in 1906 (Dale, 1906) discovered the novel substance with uterotonic properties from the extract of the human posterior pituitary, he named it "oxytocin," using Greek words meaning "quick birth". Soon after, Ott and Scott (Ott and Scott, 1910) and Schäfer and Mackenzie (Schafer and Mackenzie, 1911) described yet another "maternal" property of oxytocin - its ability to induce milk ejection during nursing.

More than a century of research on oxytocin managed to break down certain preconceptions about this nonapeptide. For a long period of time it had been considered that oxytocin was primarily produced in magnocellular neurons of paraventricular and supraoptic nuclei of hypothalamus, from where it was transported to neurohypohysis and, subsequently, released into the systemic circulation to act as a hormone. Today we know that, depending on species, oxytocin may as well be synthesized in other brain regions, such as parvocellular part of the paraventricular nucleus, bed nucleus of striaterminalis, lateral amygdala and medial preoptic area (Young and Gainer 2003). In the periphery, oxytocin was found to be produced in placenta, amnion, thymus, heart, coronary arteries and large blood vessels (Gutkowska, et al. 1997, Jankowski et al. 1998, Jankowski et al. 2000).

Apart from being a "classical hormone," oxytocin also posseses autocrine and paracrine properties and acts as a neurotransmitter and neuromodulator.

Findings of comparable concentrations of oxytocin in both male and female plasma and posterior pituitary (Gimple and Fahrenholz 2001) have made it possible to assume that the role of oxytocin in regulation of physiological functions exceeds by far previously described involvement in parturition and lactation.

Nowadays, it is well acknowledged that oxytocin plays important roles in regulating the stress response, social behaviour, hydromineral balance, growth and development, feeding behavior and energy balance, etc.

Although it is worth mentioning that oxytocin binds with low affinity to vasopressin receptors, its actions are predominantly mediated by oxytocin receptor, a member of rhodopsin-like G-protein coupled receptor family. How can the association to one type of receptor explain the pleiotropic effects of oxytocin? The possible answer lies in the promiscuous coupling of oxytocin receptor to at least three different G proteins (Gq/11, Gαi and Gαs), which further leads to activation of different second messengers and, ultimately, to variety of physiological responses (Gimpl and Fahrenholz 2001). Circumstances which determine the activation of particular signaling cascade are still waiting to be unraveled, but the activation pattern may depend on receptors` localisation in specific plasma membrane microdomains, known as lipid rafts and caveolae (Simons and Toomre 2000, Chini and Parenti 2004).

Over the last two decades, an extensive knowledge of the significance of oxytocin in cardiovascular regulation has accumulated. This paper focuses on the novel findings on the role of peripheral and central oxytocin systems in cardiovascular homeostasis.

OXYTOCIN AND CARDIOMYOGENESIS

Embryonic carcinoma cells (P19 EC) represent standard model for studying embryological development, since they possess the ability to differentiate into all three germinal layers (van der Heyden and Defize 2003). The process of differentiation from pluripotent, mouse teratocarcinoma-derived P19 EC into highly differentiated and specialized cells, including cardiomyocytes, does not happen spontaneously and requires presence of stimulatory factors. When Jankowski and collaborators detected higher levels of oxytocin in the developing compared to the adult heart, it became possible that oxytocin plays an important role in cardiac development (Jankowski et al.

2004). In fact, Paquin and collaborators (Paquin et al. 2002) were first to show that treatment of P19 cells with oxytocin as a stimulatory factor promotes formation of beating cell colonies, the process which can be prevented using oxytocin receptor antagonist. In a number of studies using different stem cell lines (Jankowski et al. 2004, Danalache et al. 2007, Hatami et al. 2007, Fathi et al. 2009) the supposed cardiomorphogenic role of oxytocin/oxytocin receptor system during development was further confirmed.

Besides having ontogenic significance, the involvement of oxytocin in the process of cardiomyogenesis may be observed in another aspect - its potential to stimulate regeneration of adult cardiac muscle. The strong belief that the heart is a postmitotic organ, with apredetermined number of terminally differentiated cardiomyocytes established at birth and without an ability for self-renewal, was challenged after the discovery of cardiac stem cells (Beltrami et al. 2003). Oxytocin promotes differentiation of stem/progenitor cells into cardiomyocytes (Matsuura K. et al. 2004, Oyama et al. 2007), and being a physiological hormone, oxytocin may be an intrinsic signal for cardiomyogenesis in the injured heart (Oyama et al. 2007).

OXYTOCIN AND THE "LITTLE BRAIN OF THE HEART"

Studies on isolated heart (Favaretto et al. 1997) and on isolated atria (Mukaddam-Daher et al. 2001) have shown that oxytocin induces negative inotropic and chronotropic effects. Since the isolated cardiac preparations are deprived of the influence of the central nervous system, the focus of attention in the understanding of the mechanisms lying behind aforementioned actions has been directed towards the intrinsic cardiac nervous system. This so called "little brain" of the heart (Armour 2008) consists of intrinsic nerve cell bodies, predominantly located in supraventricular tissues, with axons connecting to form networks that appear as ganglionic plexuses. The widespread distribution of ganglionic plexuses in the heart suggests the important role of the intrinsic cardiac nervous system in heart function. Parasympathetic postganglionic neurons in each intrinsic cardiac ganglionated plexus innervate tissues throughout the heart, and they have also been proved to contain oxytocin receptors. Once bound to its receptor, oxytocin acts in an autocrine/paracrine manner to stimulate the release of acetylcholine, which acts on atrial muscarinic receptors and induces lowering of heart rate and force of contraction (Mukaddam-Daher et al. 2001). It is well known that impaired parasympathetic function is a negative prognostic indicator in conditions such as congestive heart failure or myocardial infarction (La Rovere et al. 1998, Cole et al. 1999). By stimulating acetylcholine release and promoting vagal control in heart, oxytocin acts as a potential cardioprotective agent.

Oxytocin - The Heart Praetorian

The first physiologically relevant finding of the newly discovered oxytocin-oxytocin receptor system in heart was that oxytocin stimulates the release of atrial natriuretic peptide (Haanwinckel et al. 1995), an agent known for its numerous beneficial effects on the cardiovascular system, such as regulation of extracellular fluid volume, inhibition of sympathetic nerve activity, inhibition of renin release, etc. A few years later, the same group of researchers who discovered cardiac oxytocin, also found that oxytocin in the heart induces the release of NO (Danalche et al. 2007), another important multifaceted regulator of cardiovascular physiology. Although these first discoveries tended to present oxytocin merely as a mediator in the processes of cardioprotection, studies that followed pointed to the more direct cardioprotective role for oxytocin.

In ischemic heart disease, especially myocardial infarction, deprivation of oxygen and nutrient supply results in a series of abrupt biochemical and metabolic changes within the myocardium. The most effective therapeutic intervention for reducing acute myocardial ischemic injury and limiting the size of myocardial infarction is timely and effective myocardial reperfusion. However, myocardial reperfusion can itself induce further cardiomyocyte damage, a phenomenon known as myocardial reperfusion injury (Yellon et al. 2007). The number of *ex vivo* (Ondrejcakova et al. 2009) and *in vivo* studies (Kobayashi et al. 2009, Jankowski et al. 2010, Alizadeh et al. 2012) have repeatedly shown that oxytocin enhances cardiac recovery in an ischemia/reperfusion model of heart injury, and that oxytocin receptors have a crucial role in mediating this cardioprotective property. In oxytocin receptor-knockdown cells, oxytocin treatment resulted in elevated cell death compared to the oxytocin-untreated control cells in the ischemia/reperfusion model. It was suggested that oxytocin triggers its deleterious effect via vasopressin V1a or V2 receptors, which is in concordance with a report from Indrambarya and collaborators (Indrambaraya et al. 2009), who noticed that vasopressin infusion after an hour of ischemia increased mortality and cardiac dysfunction in mice. In addition, over-expression of V1a receptors in heart causes left ventricular dysfunction (Li et al. 2011). The most recent findings (Gonzales-

Reyes et al. 2015) suggest that the possible mechanism behind oxytocin-induced cardiomyocyte protection lies in the simultaneous activation of various signaling pathways, including ERK1/2, Pi3K/Akt and eNOS, that target mitochondria and prevent cell death.

A recent study in obese diabetic db/db mice that develop metabolic syndrome accompanied with systolic and diastolic dysfunction due to cardiomyocyte hypertrophy, fibrosis and apoptosis (Plante et al. 2015) has shown that chronic oxytocin treatment reduces cardiac oxidative stress and inflammation, and normalizes the 5-adenosine monophosphate-activated protein kinase signaling pathway, which leads to reversal of abnormal structural remodeling and improvement of cardiac function.

CONTROL OF VASCULAR RESISTANCE AND BLOOD PRESSURE BY OXYTOCIN

In contrast to broad knowledge of the contribution of vasopressin to the stability of the circulation, the effects of endogenous oxytocin on blood vessels are less evident and more controversial. It seems likely that, depending on blood vessel type and localization and/or specific physiological or pathological condition, the effects of oxytocin on blood vessels range from potent constriction to significant dilation. During pregnancy, oxytocin does not seem to play an important physiological role in the regulation of maternal vascular tone and peripheral resistance (Miller et al. 2002), while, at term, it induces strong vasoconstriction of the umbilical artery (Altemus et al. 2001). *In vitro* studies have shown that oxytocin, via vasopressin V1a receptors, induces much weaker vasoconstriction than vasopressin (Loichot C et al. 2001). However, in the condition of increased vascular tone, oxytocin may act as a vasodilator, stimulating endothelial NO release through calcium-dependent mechanisms (Russ et al. 1985, Thibonnier et al. 1999). Vasodilatatory effects of oxytocin were also described in the basilary arteries (Katusic et al. 1986).

Some studies suggest that endogenous oxytocin might be involved in the maintenance of blood pressure. Oxytocin-deficient mice show lower values of mean blood pressure compared to wild-type controls (Michelini et al. 2003). However, Nissen and collaborators (Nissen et al. 1996) demonstrated that, in lactating women, blood pressure falls in response to each nursing session.

According to the literature, both hypotensive and hypertensive effects of exogenously applied oxytocin have been reported, depending on the animal species and the route of administration.

Subcutaneous injections of oxytocin induced a short-lasting increase in blood pressure that could not be fully antagonised by oxytocin antagonist (Petterson et al. 1999). This effect is probably due to activation of vasopressin V1a receptors, since oxytocin in high doses shows affinity for vasopressin receptors.

However, this short-lasting increase in blood pressure was followed by long-term hypotension in conscious male and female rats (Petterson et al. 1996) systemically treated with oxytocin over a five-day period. Interestingly, systemically administered oxytocin in humans showed the opposite effects. Ronald Katz (1964) described transient hypotension followed by secondary rise in blood pressure after injection of synthetic oxytocin in human studies.

Neurogenic Control of the Cardiovascular System by Oxytocin

The paraventricular nucleus of the hypothalamus is recognized as the key site of autonomic cardiovascular regulation, and also the site which integrates together emotional, autonomic and endocrine responses to stress. It consists of two main parts:magnocellular and parvocellular, the former primarily involved in governing the hormonal actions of vasopressin and oxytocin and the latter divided in neuroendocrine and pre-autonomic parts.

Neurons originating from the parvocellular pre-autonomic part of the paraventricular nucleus project to autonomic nuclei crucially involved in the control of sympathetic and vagal outflow to the periphery, such as nucleus of the solitary tract, dorsal vagal nucleus, nucleus accumbens, rostroventrolateral medulla and intermediolateral column of the spinal cord (Sawchenko and Swanson 1982, Lang et al. 1982, Zerihun and Harris 1983, Hosoyaet al. 1995, Jansen et al. 1995, Hallbeck et al. 2001, Geerling et al. 2010).

In order to maintain cardiovascular homeostasis, the paraventricular nucleus adjusts sympathetic outflow according to the information delivered by the vagal nucleus of the solitary tract afferents originating from the peripheral cardiovascular receptors. Neurons located in the lateral, ventral and dorsal subdivisions of the parvocellular part of the paraventricular nucleus project to the pressor region of the rostroventrolateral medulla and sympathetic pre-ganglionic neurons in intermediolateral cell column of the thoraco-lumbar part of the spinal cord.

Parvocellular neurons affect sympathetic nerve activity in three ways: projecting directly to the intermediolateral column, indirect projections to the rostroventrolateral medulla, and through a collateral projection to both intermediolateral column and rostroventrolateral medulla (Coote et al. 1998, Pyner and Coote 1999).

As aforementioned, oxytocin is synthesized in both magnocellular and parvocellular neurons of the paraventricular nucleus and it is also well established that oxytocin receptors are normally expressed in the paraventricular nucleus (Van Leeuweet al. 1985, Freund-Mercier et al. 1987, Tribolletet al. 1988, Yoshimura et al. 1993, Adan et al. 1995) where they have an important function as a part of an endogenous autocontrol mechanism (Richard et al. 1997).

Electrophysiological studies revealed that during suckling, somato-dendritically released oxytocin stimulates oxytocin receptors on magnocelluar neurons in the paraventricular nucleus to increase the basal firing rate and establish a periodic bursting activity pattern. The underlying mechanisms involve priming of oxytocin neurons and release of calcium from IP3-sensitive intracellular stores (Inenaga and Yamashita 1986, Moos and Richard 1989, Richard et al. 1997, Ludwig and Leng 2006). Anatomical and electrophysiological studies indicate a potentially important role of the paraventricular oxytocinergic system in autonomic cardiovascular control, since approximately 40% of the spinally projecting paraventricular nucleus neurons contain mRNA for oxytocin (Pyner 2009). According to existing knowledge, the oxytocin system, together with vasopressin, dopamine and angiotensin II in the paraventricular nucleus, is found to selectively modulate tonic paraventricular nucleus signal in autonomic cardiovascular control (Pyner 2009). By combining the advantages of spectral analysis technique that gives an insight into cardio-respiratory control (Japundžić-Žigon 1998) with over-expression of oxytocin receptors and microinfusions of selective oxytocin antagonist in the paraventricular nucleus, we provided new evidence on the role of oxytocin system in central regulation of cardiovascular system (Lozic et al. 2014). This study suggests that although oxytocin receptors in the paraventricular nucleus have no tonic influence on mean levels of blood pressure and heart rate under baseline physiological conditions, they activate downstream signaling pathways in neighbouring cells leading to neurotransmitter release in brainstem targets that increase baroreceptor reflex sensitivity and reduce blood pressure variability, markers correlated with positive clinical outcome in cardiovascular diseases (Mancia et al. 1994, Narkiewicz and Grassi 2008).

Oxytocin and the Cardiovascular Response to Stress

It is well known that psychological stress elicits significant changes in the hypothalamo-pituitary-adrenal axis as well as in autonomic balance, which can negatively affect the cardiovascular system (Brotman et al. 2007). Exposure to chronic stress is the risk factor in etiopathogenesis of cardiovascular disease, while acute stress can trigger myocardial infarction and sudden death (Japundzic-Zigon 2013).

A significant number of animal studies have suggested that oxytocin acts as an anti-stress hormone (Grippo et al. 2009, Lee et al. 2005, Windle et al. 1997, 2004). For instance, oxytocin is found to decrease cardiovascular response to isolation in monogamous praire voles, that represent an excellent model for studying social behavior (Grippo et al. 2009). Oxytocin blunts restraint-induced hypothalamo-pituitary axes activation (Windle et al. 1997, 2004) by reducing stress-induced elevation of plasma corticosterone in rats, and promotes social interactions (Lee et al. 2005). Rats treated centrally with oxytocin showed a higher proportion of open-arm entry and spent more time in the open arms in the elevated plus maze, expressing signs of reduced anxiety-like behaviour (Windleet al. 1997). In oxytocin knock-out mice, Bernatova and co-workers (2004) described accentuated blood pressure and corticosterone responses during exposure to acute stress, while the application of oxytocin antisense oligonucleotides into the paraventricular nucleus (Callahan et al. 1989) attenuated heart rate response to stress. In line with previous findings, Wsol and colleagues (2008) reported that the central application of oxytocin receptor antagonist enhanced blood pressure and heart rate increases caused by environmental stress. Results from our study (Lozic et al. 2014) show that over-expression of oxytocin receptors in the paraventricular nucleus buffered stress-induced blood pressure variability response and favored vagal control of the heart. The increase of vagal influence to the heart during stress was reported to be useful in protecting the heart against sympathetic over-stimulation (Bracket al. 2012). This protective effect of the vagus is lifesaving during cardiac ischemia, when sympathetic over-stimulation triggers life threatening arrhythmias and sudden death. This assumption is further supported by the work of Wsol and associates (2009). In a series of ellegant experiments in rats that survived myocardial infarction after ligation of coronary arteries, the authors reported reduced survival due to the failure of brain oxytocin to attenuate cardiovascular response to stress.

Clinical findings in humans also support a role for oxytocin as an anti-stress hormone. Altemus and collaborators (Altemus et al. 2001) reported that lactating women have greater parasympathetic control of the heart, and Grewen and Light (2011) found that plasma oxytocin in lactating women is correlated with lower cardiovascular reactivity to stress.

THE ROLE OF OXYTOCIN IN THE ADJUSTMENT OF MATERNAL CIRCULATION TO PREGNANCY

Pregnancy is a physiological state where hemodynamic changes occur to meet the oxygen and nutritional requirements of the growing uterus and developing fetus. It is well established that oxytocin, along with vasopressin, is involved in the mediation of the hypervolaemic and hyponatremic state in pregnancy (Thibonnier et al. 2001). By the end of pregnancy blood volume is increased by 55% and plasma sodium concentration and osmolality are decreased by 4% (Thibonnier et al. 2001). This new hydromineral balance associated with pregnancy is induced by corpora lutea derived relaxin that stimulates vasopressin release and drinking behavior (Gray et al. 2012) and concomitant quiescence of oxytocin neurons (Krause et al. 2012). The quiescence of oxytocin neurons to osmotic stimuli during pregnancy might be induced by allepregnenolon, a progesterone neurosteroid metabolite (Tribollet et al. 1988, Gimpl and Fahrenholz 2001). As a consequence of hypervolaemia, maternal cardiac output increases and this is accompanied by gradual increase of myocardial contractility provoking mild left ventricular hypertrophy by the end of pregnancy (Hasser et al. 1997). At the same time, remodeling of cardiovascular neurogenic control occurs and baro-receptor reflex sensitivity is decreased (Barberis et al. 1998), as well as peripheral resistance and blood pressure. Moreover, the responsiveness of maternal vessels to pressor agents is reduced, while the responsiveness to vasodilators is enhanced.

CONCLUSION

Extensive research during the last two decades has persuaded us that oxytocin indeed is "the great facilitator of life" (Lee et al. 2009). From the aspect of phylogeny, oxytocin serves to propagate species as it is crucially entwined in the act of birth, sexuality and parenthood. Ontogenetically, oxytocin stimulates embryological development, promotes cell differentiation and organogenesis. Its role in the protection of cardiovascular system is multilayered-oxytocin synthesized in heart acts directly on cardiomyocytes, promoting their survival and regeneration, while centrally produced oxytocin protects cardiovascular health by moderating the stress response. Further research on oxytocin and the cardiovascular system should contribute to the discovery of new targets for drug discovery and should bring advances in stem cell therapy approach.

REFERENCES

Adan R. A. H., Van Leeuwen F. W., Sonnemans M. A. F., Brouns M., Hoffman G., Verbalis J. G., Burbach J. P. (1995). Rat oxytocin receptor in brain, pituitary, mammary gland, and uterus: partial sequence and immunocytochemical localization. *Endocrinology,* 136(9):4022 - 4028.

Alizadeh A. M., Faghihi M., Khori V., Sohanaki H., Pourkhalili K., Mohammadghasemi F., Mohsenikia M. (2012). Oxytocin protects cardiomyocytes from apoptosis induced by ischemia-reperfusion in rat heart: role of mitochondrial ATP-dependent potassium channel and permeability transition pore. *Peptides,* 36(1):71 - 7.

Altemus M., Redwine L. S., Leong Y. M., Frye C. A., Porges S. W., Carter C. S. (2001). Responses to laboratory psychosocial stress in postpartum women. *Psychosomatic medicine*, 63(5):814 - 821.

Armour J. A. (2008). Potential clinical relevance of the 'little brain' on the mammalian heart. *Exp. Physiol.*, 93(2):165 - 76.

Barberis C., Mouillac B., Durroux T. (1998). Structural bases of vasopressin oxytocin receptor function. *J. Endocrinol.*, 156:223 – 229.

Beltrami A. P., Barlucchi L., Torella D., Baker M., Limana F., Chimenti S., Kasahara H., Rota M., Musso E., Urbanek K., Leri A., Kajstura J., Nadal-Ginard B., Anversa P. (2003). Adult cardiac stem cells are multipotent and support myocardial regeneration. *Cell,* 114(6):763 - 76.

Benarroch E. E. (2005). Paraventricular nucleus, stress response, and cardiovascular disease. *Clin. Auton. Res.*, 15:254 - 263.

Bernatova I., Rigatto K. V., Ke M. P., Morris M. (2004). Stress-induced pressor and corticosterone response in oxytocine-deficient mice. *Exp. Physiol.*, 89:549 - 557.

Brack K. E., Winter J., Ng G. A. (2012). Mechanisms underlying the autonomic modulation of ventricular fibrillation initiation-tentative prophylactic properties of vagus nerve stimulation on malignant arrhythmia in heart failure. *Heart Fail Rev.*, 18:389 – 408.

Callahan M. F., Kirby R. F., Cunningham T., Eskridge-Sloop S. L., Johnson A. K., McCarthy R. et al. (1989). Central oxytocin systems may mediate a cardiovascular response to acute stress in rats. *Am. J. Physiol. Heart Circ. Physiol.*, 256 (25):H1369-H1377.

Chini B., Parenti M. (2004). G-protein coupled receptors in lipid rafts and caveolae: how, when and why do they go there? *Journal of Molecular Endocrinology*, 32:325 - 338.

Cole C. R., Blackstone E. H., Pashkow F. J., Snader C. E., Lauer M. S. (1999). Heart-rate recovery immediately after exercise as a predictor of mortality. *N Engl. J. Med.*, 28;341(18):1351 - 1357.

Dale HH (1906). On some physiological actions of ergot. *J Physiol* 34(3):163-206.

Danalache B. A., Paquin J., Donghao W., Grygorczyk R., Moore J. C., Mummery C. L., Gutkowska J., Jankowski M. (2007). Nitric oxide signaling in oxytocin-mediated cardiomyogenesis. *Stem Cells*, 25(3):679-688.

Fathi F., Murasawa S., Hasegawa S., Asahara T., Kermani A. J., Mowla S. J. (2009). Cardiac differentiation of P19CL6 cells by oxytocin. *Int. J. Cardiol.*, 134(1):75-81.

Favaretto A. L., Ballejo G. O., Albuquerque-Araújo W. I., Gutkowska J., Antunes-Rodrigues J., McCann S. M. (1997). Oxytocin releases atrial natriuretic peptide from rat atria in vitro that exerts negative inotropic and chronotropic action. *Peptides,* 18(9):1377 - 1381.

Gimpl G., Fahrenholz F. (2001). The oxytocin receptor system: structure, function, and regulation. *Physiol. Rev.*, 81(2): 629 - 683.

Gray M., Innala L., Viau V. (2012). Central vasopressin V1A receptor blockade impedes hypothalamic-pituitary-adrenal habituation to repeated restraint stress exposure in adult male rats. *Neuro-psychopharmacology*, 37(12):2712 – 2719.

Grippo A. J., Trahanas D. M., Zimmerman R. R. 2nd, Porges S. W., Carter C. S. (2009). Oxytocin protects against negative behavioral and autonomic consequences of long-term social isolation. *Psychoneuroendocrinology*, 34(10):1542 - 1553.

Gutkowska J., Antunes-Rodrigues J., McCann S. M. (1997). Atrial natriuretic peptide in brain and pituitary gland. *Physiol. Rev.*, 77:465 - 515.

Gutkowska J., Jankowski M., Lambert C., Mukaddam-Daher S., Zingg H. H., McCann S. M. (1997). Oxytocin releases atrial natriuretic peptide by combining with oxytocin receptors in the heart. *Proc. Natl. Acad. Sci. U S A*, 94(21):11704 - 11709.

Haanwinckel M. A., Elias L. K., Favaretto A. L., Gutkowska J., McCann S. M., Antunes-Rodrigues J. (1995). Oxytocin mediates atrial natriuretic peptide release and natriuresis after volume expansion in the rat. *Proc. Natl. Acad. Sci. U S A*, 15;92(17):7902 - 7906.

Hatami L., Valojerdi M. R., Mowla S. J. (2007). Effects of oxytocin on cardiomyocyte differentiation from mouse embryonic stem cells. *Int. J. Cardiol.*, 117(1):80 - 89.

Inenaga K., Yamashita H. (1986). Excitation of neurones in the rat paraventricular nucleus in vitro by vasopressin and oxytocin. *J. Physiol.*, 370:165 - 180.

Jankowski M., Hajjar F., Kawas S. A., Mukaddam-Daher S., Hoffman G., McCann S. M., et al. Rat heart: a site of oxytocin production and action (1998). *Proc. Natl. Acad. Sci. U S A*, 95:14558 - 14563.

Jankowski M., Wang D., Hajjar F., Mukaddam-Daher S., McCann S. M., Gutkowska J. (2000). Oxytocin and its receptors are synthesized in the rat vasculature. *Proc. Natl. Acad. Sci. U S A*, 97:6207 - 6211.

Jankowski M., Danalache B., Wang D., Bhat P., Hajjar F., Marcinkiewicz M., et al. (2004). Oxytocin in cardiac ontogeny. *Proc. Natl. Acad. Sci. U S A*, 101:13074 - 13079.

Japundzic-Zigon N. (1998). Physiological mechanisms in regulation of blood pressure fast frequency variations. *Clin. Exp. Hypertens.*, 20(4):359 - 388.

Japundzic-Zigon N. (2013). Vasopressin and oxytocin in control of the cardiovascular system. *Curr. Neuropharmacol.*, 11:218-230.

Katusic Z. S., Shepherd J. T., Vanhoutte P. M. (1986). Oxytocin causes endothelium-dependent relaxations of canine basilar arteries by activating V1-vasopressinergic receptors. *J. Pharmacol. Exp. Ther.*, 236(1):166 - 170.

La Rovere M. T., Bigger J. T. Jr., Marcus F. I., Mortara A., Schwartz P. J. (1998). Baroreflex sensitivity and heart-rate variability in prediction of total cardiac mortality after myocardial infarction. ATRAMI (Autonomic Tone and Reflexes After Myocardial Infarction) Investigators. *Lancet*, 351(9101):478 - 84.

Lang R. E., Heil J., Ganten D., Hermann K., Rascher W., Unger Th. (1983). Effects of lesions in the paraventricular nucleus of the hypothalamus on vasopressin and oxytocin contents in brainstem and spinal cord of rat. *Brain Res.*, 260:326 - 329.

Lee, H.-J., Macbeth, A. H., Pagani, J., and Young, W. S. (2009). Oxytocin: the Great Facilitator of Life. *Progress in Neurobiology*, 88(2):127 – 151.

Lee P. R., Brady D. L., Shapiro R. A., Dorsa D. M., Koenig J. I. (2005). Social interaction deficits caused by chronic phencyclidine administration are reversed by oxytocin. *Neuropsychopharmacology,* 30(10):1883 - 1894.

Loichot C., Krieger J. P., De Jong W., Nisato D., Imbs J. L., Barthelmebs M. (2001). High concentrations of oxytocin cause vasoconstriction by activating vasopressin V1A receptors in the isolated perfused rat kidney. *Naunyn Schmiedebergs Arch. Pharmacol.,* 363(4):369 - 375.

Lozić M., Greenwood M., Šarenac O., Martin A., Hindmarch C., Tasić T., Paton J., Murphy D., Japundžić-Žigon N. (2014). Overexpression of oxytocin receptors in the hypothalamic PVN increases baroreceptor reflex sensitivity and buffers BP variability in conscious rats. *Br. J. Pharmacol.,* 171(19):4385 - 4398.

Ludwig M., Leng G. (2006). Dendritic peptide release and peptide-dependent behaviours. *Nat. Rev. Neurosci.,* 7:126 - 136.

Mancia G., Frattola A., Parati G., Santucciu C., Ulian L. (1994). Blood pressure variability and organ damage. *J. Cardiovasc. Pharmacol.,* 24 (Suppl. A): S6-S11.

Michelini L. C., Marcelo M. C., Amico J., Morris M. (2003). Oxytocinergic regulation of cardiovascular function: studies in oxytocin-deficient mice. *Am. J. Physiol. Heart Circ. Physiol.,* 284:H2269-H2276.

Moos F., Richard P. (1989). Paraventricular and supraoptic bursting oxytocin cells in rat are locally regulated by oxytocin and functionally related. *J. Physiol.,* 408:1 - 18.

Mukaddam-Daher S., Yin Y. L., Roy J., Gutkowska J., Cardinal R. (2001). Negative inotropic and chronotropic effects of oxytocin. *Hypertension,* 38(2):292 - 296.

Nishioka T., Anselmo-Franci J. A., Li P., Callahan M. F., Morris M. (1998). Stress increases oxytocin release within the hypothalamic paraventricular nucleus. *Brain Res.,* 781:57 - 61.

Nissen E., Uvnäs-Moberg K., Svensson K., Stock S., Widström A. M., Winberg J. (1996). Different patterns of oxytocin, prolactin but not cortisol release during breastfeeding in women delivered by caesarean section or by the vaginal route. *Early Hum. Dev.,* 45(1-2):103 - 18.

Narkiewicz K., Grassi G. (2008). Imapiredbaroreflex sensitivity as a potential marker of cardiovascular risk in hypertension. *J. Hypertens.,* 26:1303-1304.

Ondrejcakova M., Ravingerova T., Bakos J., Pancza D., Jezova D. (2009). Oxytocin exerts protective effects on in vitro myocardial injury induced by ischemia and reperfusion. *Can. J. Physiol. Pharmacol.,* 87:137 - 142.

Ott I, Scott JC (1910). The action of infundibulum upon mammary secretion. *Proc Soc Exp Biol* 8:48-49.

Oyama T., Nagai T., Wada H., Naito A. T., Matsuura K., Iwanaga K., et al. (2007). Cardiac side population cells have a potential to migrate and differentiate into cardiomyocytes in vitro and in vivo. *J. Cell Biol.,* 176:329 - 341.

Paquin J., Danalache B. A., Jankowski M., McCann S. M., Gutkowska J. (2002). Oxytocin induces differentiation of P19 embryonic stem cells to cardiomyocytes. *Proc. Natl. Acad. Sci. U S A,* 99(14):9550 - 9555.

Petersson M., Unväs-Moberg K. (2007). Effects of an acute stressor on blood pressure and heart rate in rats pretreated with intracerebroventricular oxytocin injections. *Psychoneuroendocrinology,* 32:959 - 965.

Plante E., Menaouar A., Danalache B. A., Yip D., Broderick T. L., Chiasson J. L., Jankowski M., Gutkowska J. (2015). Oxytocin treatment prevents the cardiomyopathy observed in obese diabetic male db/db mice. *Endocrinology,* 156(4):1416 - 1428.

Pyner S. (2009). Neurochemistry of the paraventricular nucleus of the hypothalamus: implications for cardiovascular regulation. *J. of Chemical Neuroanatomy,* 38:197 - 208.

Richard P., Moos F., Dayanithi G., Gouzènes L., Sabatier N. (1997). Rhythmic activities of hypothalamic magnocellular neurons: autocontrol mechanisms. *Biol. Cell,* 89:555 - 560.

Schafer EA, Mackenzie K. (1911). The action of animal extracts on milk secretion. Proc Royal Soc London Series B 84(568):16-22.

Simons K., Toomre D. (2000). Lipid rafts and signal transduction. *Nat. Rev. Mol. Cell Bioll.,* 1(1):31 – 39.

Swanson L. W. (1995). Mapping the human brain: past, present and future. *Trends Neurosci.,* 18:471 - 474.

Thibonnier M., Conarty D. M., Preston J. A., Plesnicher C. L., Dweik R. A., Erzurum S. C. (1999). Human vascular endothelial cells express oxytocin receptors. *Endocrinology,* 140(3):1301 - 1309.

Thibonnier M., Coles P., Thibonnier A., Shoham M. (2001). The basic and clinical pharmacology of nonpeptide vasopressin receptor antagonists. *Annu. Rev. Pharmacol. Toxicol.,* 41:175–202.

Tribollet E., Barberis C., Jard S., Dubois-Dauphin M., Dreifuss J. J. (1988). Localization and pharmacological characterization of high affinity binding sites for vasopressin and oxytocin in the rat brain by light microscopic autoradiography. *Brain Res.,* 442:105–118.

van der Heyden M. A., Defize L. H. (2003). Twenty one years of P19 cells: what an embryonal carcinoma cell line taught us about cardiomyocyte differentiation. *Cardiovasc. Res.*, 58(2):292 - 302.

Windle R. J., Shanks N., Lightman S. L., Ingram C. D. (1997). Central oxytocin administration reduces stress-induced corticosterone release and anxiety behavior in rats. *Endocrinology*, 138(7):2829 - 2834.

Windle R. J., Kershaw Y. M., Shanks N., Wood S. A., Lightman S. L., Ingram C. D. (2004). Oxytocin attenuates stress-induced c-fos mRNA expression in specific forebrain regions associated with modulation of hypothalamo-pituitary-adrenal activity. *J. Neurosci.*, 24(12):2974 - 2982.

Wsol A., Cudnoch-Jedrzejewska A., Szczepanska-Sadowska E., Kowalewski S., Puchalska L. (2008). Oxytocin in the cardiovascular responses to stress. *J. Physiol. Pharmacol.*, 59(Suppl. 8):123 - 127.

Wsol A., Cudnoch-Jedrzejewska A., Szczepanska-Sadowska E., Kowalewski S., Dobruch J. (2009). Central oxytocin modulation of acute stress-induced cardiovascular response after myocardial infarction in the rat. *Stress,* 12(6):517-525.

Yellon D. M., Hausenloy D. J. (2007). Myocardial reperfusion injury. *N Engl. J. Med.*, 357(11):1121 – 1135.

Young III W. S., Gainer H. (2003). Transgenesis and the study of expression, cellular targeting and function of oxytocin, vasopressin and their receptors. *Neuroendocrinology*, 78:185 - 203.

Zerihun L., Harris M. (1983). An electrophysiological analysis of caudally projecting neurones from the hypothalamic paraventricular nucleus in the rat. *Brain Res.*, 261:13 - 20.

In: Advances in Oxytocin Research
Editor: Evan Anderson

ISBN: 978-1-63483-991-4
© 2015 Nova Science Publishers, Inc.

Chapter 3

THE INFLUENCE OF OXYTOCIN ON FACE-PROCESSING SKILLS

Rachel Bennetts, PhD, and Sarah Bate[*]*, PhD*
Bournemouth University, Poole, UK

ABSTRACT

Oxytocin is a nonapeptide that plays a fundamental role in social cognition, and recent evidence demonstrates that intranasal inhalation of the hormone improves face-processing. For example, a number of studies have found that people are significantly better at recognising facial expressions of emotion following oxytocin inhalation; and there is some evidence that oxytocin improves facial identity memory. However, the benefits of oxytocin for facial identity recognition are somewhat inconsistent, with some studies finding effects under limited circumstances (e.g., only when the faces show certain emotional expressions, or only amongst certain groups of participants), and others finding selective, null or negative effects. Notably, some evidence suggests that oxytocin is not always facilitative, raising the possibility that it may only improve face recognition performance under certain conditions. Currently, the mechanisms underpinning the link between oxytocin and improved face-processing are unclear. Oxytocin appears to improve face encoding (as opposed to retrieval of a face from memory), but it is unclear whether this effect arises due to changes in face perception or later encoding or memory consolidation processes, nor why

---

[*] E-mail: sbate@bournemouth.ac.uk.

this effect occurs. It is possible that oxytocin acts upon the core face-processing system, and some neuroimaging evidence supports this possibility. Alternatively, oxytocin may affect face-processing indirectly by modulating social salience, approach-withdrawal motivation, or general levels of anxiety. This commentary will summarise the available literature on this issue, attempting to clarify when oxytocin does and does not benefit face recognition, and the mechanisms that underpin this effect.

INTRODUCTION

The ability to recognise faces and interpret their emotional expression is a vital aspect of human social cognition. However, face recognition abilities vary widely throughout the population (Bowles et al. 2009, Wilmer et al. 2012), and difficulties with face recognition occur in a number of neuropsychological disorders (e.g., autism spectrum disorder, Weigelt, Koldewyn and Kanwisher 2012, prosopagnosia, Susilo and Duchaine 2013). Consequently, there is much interest in developing techniques or interventions to improve face-processing. One such intervention is oxytocin: a nonapeptide that plays a fundamental role in social cognition (Heinrichs, von Dawens and Domes 2009).

Recent evidence demonstrates that oxytocin can influence face-processing – for example, natural variation in oxytocin receptor genes is associated with performance on face memory and facial expression recognition tests (Skuse et al. 2014). Furthermore, intranasal inhalation of synthetic oxytocin can lead to enhanced performance on tasks such as facial expression recognition and memory for facial identity (Guastella, Mitchell and Mathews 2008a, Rimmele, Hediger, Heinrichs and Klaver 2009, Savaskan, Ehrardt, Schulz, Walter and Schachinger 2008, Shahrestani, Kemp and Guastella 2013, van Ijzendoorn and Bakermans-Kranenburg 2012). However, the benefits of oxytocin for face-processing are somewhat inconsistent, and the mechanisms underlying these benefits are unclear. If oxytocin is to be used as a therapeutic intervention to improve face-processing abilities, it is crucial that we understand when and why it brings about improvements. Several recent reviews and meta-analyses have examined the effects of oxytocin on facial expression recognition (see Evans, Dal Monte, Noble and Averbeck 2014, Guastella and MacLeod 2012, Shahrestani et al. 2013, van Izjendoorn and Bakersman-Kranenburg 2012), but there has been little work that attempts to summarise and critically examine the research on oxytocin and facial identity processing. First, this commentary reviews the evidence suggesting that oxytocin can improve face memory.

Second, it attempts to situate behavioural findings within a cognitive model of face recognition, to examine how oxytocin improves memory for faces. Finally, the main theoretical positions that attempt to account for the social and cognitive effects of oxytocin on face recognition are discussed.

DOES OXYTOCIN IMPROVE MEMORY FOR FACES?

A number of studies have examined whether oxytocin improves memory for individual faces. Most studies have used similar procedures: participants inhale 20-24IU of oxytocin or placebo spray shortly before (or occasionally just after) exposure to a number of unfamiliar faces. Subsequently (between 30 minutes and one day after inhalation), participants are asked to identify faces from an array of previously seen and newly presented faces. Despite this relatively uniform procedure, studies have found variable results. Only one of nine studies has found unequivocally positive effects: Rimmele et al. (2009) found a significant benefit of oxytocin for face memory (positive, neutral, and negative expressive faces), but not memory for non-social stimuli (houses, scenes, and objects). Other studies have found more equivocal results. For example, Guastella et al. (2008a) found that participants who had inhaled oxytocin were significantly better than those who had inhaled a placebo when asked to identify previously seen happy faces, but the effect of oxytocin was not present for neutral or angry faces. Similarly, Herzmann, Bird, Freeman, and Curran (2013) found a stronger effect of oxytocin for memory of happy faces than neutral faces. In contrast, Savaskan et al. (2008) found that inhalation of oxytocin following exposure to faces significantly improved performance for neutral and angry, but not happy, faces.

Similarly contrasting results have been found when investigating the role of oxytocin in recognition of other-race faces: Herzmann et al. (2013) found that oxytocin improved memory for both own- and other-race faces, whereas Blandon-Gitlin, Pezdek, Saldivar, and Steelman (2014) found that oxytocin significantly improved recognition of other-race (but not own-race) faces.

The pattern of results for oxytocin and face memory is further complicated by studies that have found null or negative effects. Di Simplicio et al. (2009) found no effect of oxytocin using a standardised test of face memory, the Cambridge Face Memory Test (CFMT, Duchaine and Nakayama 2006) (see also Bate et al. 2014, control participant results). Similarly, Bate, Bennetts, Bindemann, Udale, and Bussant (2015) found no effect on overall accuracy of face recognition when participants were required to select a target face from an array of 10 faces. While participants in this study were significantly better at identifying the correct face under oxytocin conditions, they were also more likely to select incorrect faces (or incorrectly state that the target face was present when it was not) – as such, there was no overall beneficial effect of oxytocin, merely a shift in participants' bias to respond "present." Finally, Herzmann et al. (2012) found a significant negative effect of oxytocin for memory of faces and houses.

One study has found that the effects of oxytocin on face-processing may also vary between individuals. Bate et al. (2014) found a significant benefit of oxytocin for face memory and face-matching in participants with developmental prosopagnosia, a severe deficit of face recognition. However, in line with Di Simplicio et al. (2009), there was no benefit of oxytocin for face memory or face matching in control participants with typical face recognition abilities.

In sum, three studies have found null or negative effects of oxytocin on face memory, while six studies have found some improvement in face recognition memory (although the exact details of when these effects appear vary significantly between studies). How can we account for such variability in the results? One potential explanation is variation in methodological factors. As noted above, many of the studies in this field use highly similar methods. However, three of the studies that reported null or variable findings departed from the methodologies described above. Studies using the CFMT (Di Simplicio et al. 2009, Bate et al. 2014) and the 1-in-10 matching task (Bate et al. 2015) presented the learning and test faces in a single session, around 45 minutes after inhalation of oxytocin. As such, it is possible that these methodological variations can account for the null or negative results – perhaps oxytocin is only beneficial in longer-term memory tasks (e.g., tasks in which there is a delay of at least 30 minutes between encoding and retrieval). While this may account for some of the discrepant findings, it is still puzzling that studies with almost identical methodologies (e.g., Guastella et al. 2008, Rimmele et al. 2009, Savaskan et al. 2008, Blandon-Gitlin et al. 2014, Herzmann et al. 2013) have found such variable effects of oxytocin on face memory. Guastella and MacLeod (2012) suggested that encoding instructions may orient participants to facial expression (or perhaps race of face) to different degrees, making the emotion of the faces more salient in some experiments than others. However, to date this possibility has not been formally tested.

Based on the results from Bate et al. (2014), another potential way to reconcile these conflicting results is to consider the idea that oxytocin may only be useful when face recognition is impaired or sub-optimal. Bartz et al. (2011) reviewed the effects of oxytocin on a wide range of social cognition tasks, and concluded that they were almost always modulated by personal or situational factors. Weisman and Feldman (2013) extended this idea by proposing a model in which individual characteristics and situational variables interacted with oxytocin to produce changes in behaviour and cognition when competence at a task was low, but no changes (i.e., a ceiling effect) when competence is normal or high. Studies on other aspects of social cognition involving face-processing support this interactionist approach, both on an intra- and inter-individual level. For example, on an intra-individual level, Domes, Heinrichs, Michel, Berger, and Herpetz (2007) found a significant effect of oxytocin for emotional expression recognition, which was enhanced for difficult test items; similarly, oxytocin lowers the intensity level necessary to identify an emotional expression (Lischke et al. 2012). On the inter-individual level, Bartz et al. (2010) found that inhalation of oxytocin significantly increased empathic accuracy in individuals with high autistic traits, but not those with low autistic traits.

It is possible, therefore, that the effects of oxytocin may also vary depending on whether we process and remember particular faces (e.g., particular facial expressions or other-race faces) poorly. This is in line with studies of the other-race effect: Blandon-Gitlin et al. (2014) found that the beneficial effects of oxytocin were isolated to other-race faces, which were remembered more poorly than own-race faces in the placebo condition (see Young, Hugenberg, Bernstein, and Sacco, 2012, for a review of the other-race effect). It is possible that Herzmann et al. (2013) failed to find a comparable effect simply because the other-race faces in their study were not processed as poorly as those in Blandon-Gitlin et al.'s (2014) study. There is also some evidence that facial expressions can interact with face memory in the absence of oxytocin, with happy faces remembered more accurately than angry faces (e.g., D'Argembeau, Van der Linden, Comblain and Etienne 2003). It may be that oxytocin ameliorates this effect, leading to better recognition of angry faces (Savaskan et al., 2009). Once again, though, it is unclear why the same effects would be observed for happy faces (Guastella et al. 2008a, Herzmann et al. 2012), which are generally remembered quite well. Overall, the majority of research suggests that oxytocin does improve face recognition, although the effects are nebulous and may vary depending on the facial stimuli being recognised, as well as the characteristics of the perceiver. The following sections examine this phenomenon in more detail, examining how and why this improvement might arise.

HOW DOES OXYTOCIN IMPROVE MEMORY FOR FACES?

In order to fully understand the conditions under which oxytocin may be useful as a therapeutic intervention, it is vital to examine *how* this improvement comes about.

Recognising an individual face is a complicated, multi-stage perceptual and cognitive process. First, one must extract relevant perceptual information from the face, a process involving complex, fine-grained discriminations of individual features and their spatial relationships (Maurer, Le Grand and Mondloch 2002). For unfamiliar faces, this information must then be encoded and stored as a unique identity. Subsequently, one must link incoming visual information to this stored visual representation – success in this stage leads to a sense of familiarity with the face, without necessarily knowing who the person is or why they are familiar. Finally, this visual representation must be linked to semantic information about the person (Bruce and Young 1986). Changes at any stage of this process may modulate face recognition abilities – for example, improved perceptual processing may lead to better stored visual representations; improved encoding of visual representations may make it easier to say that an individual is familiar (or, conversely, say that an unfamiliar face has not been seen before); or improved links between visual and semantic stores may make it easier to recollect exact details about a person.

Research using the remember/know paradigm (Jacoby 1991) gives some hints as to which stages of the face-processing system are affected by oxytocin. The majority of studies that have distinguished between "familiarity" and "recollection" (i.e., accessing just the visual representation versus accessing specific semantic information about the face) have found that oxytocin affects familiarity judgements (e.g., Guastella et al. 2008a, Herzmann et al. 2013, Rimmele et al. 2009), whereas effects on recollection judgements are mixed (positive in Guastella et al. 2008, no effect in Rimmele et al. 2009, negative in Herzmann et al. 2012, 2013), suggesting that oxytocin has a relatively consistent, positive effect for accessing and matching stored visual representations, but not necessarily linking these representations to semantic information. In other words, oxytocin appears to strengthen encoding of or access to visual representations, but not necessarily semantic representations.

This still leaves open the question of *how* oxytocin affects visual representations of faces: via better initial perceptual processes, enhanced encoding or consolidation of visual information, or more effective retrieval of stored visual representations. Several studies have found that behavioural effects of oxytocin emerge when it has been administered prior to or just after the learning phase of face recognition experiments (e.g., Guastella et al. 2008a, Rimmele et al. 2008, Savaskan et al. 2009). Since the effects of oxytocin are short-lived (Striepens et al. 2013), and many studies have tested retrieval up to 24 hours after inhalation (e.g., Guastella et al. 2008a, Herzmann et al. 2012, Rimmele et al. 2008), it is unlikely that oxytocin affected retrieval memory (note that this does not imply that oxytocin cannot improve retrieval of faces from memory, just that the experimental procedures used to date have not tested this possibility). Therefore, it is likely that oxytocin affected earlier stages of face recognition, such as perception, encoding, or memory consolidation. However, studies examining the distinction between perception/encoding and consolidation have found mixed results. Savaskan et al. (2008) found a significant benefit of oxytocin when administered after encoding, suggesting it modulated memory consolidation processes. In contrast, the modulating effect of oxytocin on the other-race effect was present when oxytocin was administered before encoding, but disappeared when oxytocin was administered after encoding, leading Blandon-Gitlin et al. (2014) to conclude that the benefits of oxytocin are confined to the initial perceptual or encoding phases.

At present it is difficult to say whether these findings reflect changes in perception or encoding processes. It is quite difficult to disentangle changes in perception from changes in encoding, as there has been very little research with oxytocin focussing on identity perception (as opposed to memory) and no studies have directly assessed the visual representation of faces encoded with oxytocin. Studies examining face-processing in non-identity based tasks hint that oxytocin may have an effect on face perception: for example, there is a large amount of evidence that oxytocin improves emotional expression recognition (Shahrestani et al. 2013, van Izjendoorn and Bakersman-Kranenburg 2012). Furthermore, some studies using eye-tracking have found that oxytocin increases time spent examining the eye region of the face (generally thought to be an important region for face-processing, Peterson and Eckstein 2012) (Gamer et al. 2010, Guastella et al. 2008b, cf. Lischke et al. 2012), suggesting that oxytocin may influence visual attention to faces. However, the only direct tests of facial identity perception come from face matching tasks used by Bate et al. (2014) and Bate et al. (2015). Neither found an overall benefit of oxytocin in participants with typical face recognition abilities, suggesting either a ceiling effect for typical participants or a later locus of effect for oxytocin. These behavioural results are supported by Herzmann et al. (2013), who examined EEG responses during their face recognition task, and failed to find any effect of oxytocin on ERPs related to perceptual processing of faces. These authors suggested that the effects of oxytocin were not perceptual, but related to the encoding stages of face recognition.

Overall, evidence from behavioural studies suggests that the effects of oxytocin arise because individuals are better able to access stored visual representations of faces, probably due to improvements in early stages of face memory. However, there have been conflicting results when attempting to assess the role of oxytocin in face perception and encoding/consolidation. Once again, these differences may be a reflection of oxytocin interacting with specific situational factors, including aspects of the cognitive process that are not already at ceiling (i.e., perception or encoding of other-race faces, consolidation of a large number of faces, or perception of faces in prosopagnosics).

Finally, it is important to ask whether the effects of oxytocin are specific to face memory, or whether they reflect more general modulation of mnemonic processes. Research in rodents suggests that oxytocin plays an important role in memory, but that its effects are specific to social stimuli: for example, mice bred to lack the oxytocin peptide fail to recognise their familiar conspecifics, but perform normally on non-social memory tasks (e.g., water mazes, food searches) (Ferguson, Young and Insel 2002). Studies into the effects of oxytocin on humans point to a similar conclusion: early research into the cognitive effects of oxytocin found negligible or even negative effects on memory for words and other non-social stimuli such as Rey figures (Fehm-Wolfsdorf and Born 1991, Heinrichs, Meinlschmidt, Wippich, Ehlert and Hellhammer 2004), in contrast to the apparent benefit for faces. Rimmele et al. (2009) produced one of the most convincing demonstrations of selective effects of oxytocin on social memory. As mentioned above, they found that intranasal inhalation of oxytocin selectively increased recognition of faces, but not non-social images (houses, landscapes, and objects).

However, decades of research into face-processing has indicated that faces differ from other visual stimuli in a number of ways apart from their social nature. Faces are a highly complex, naturally varying stimulus that share an overall configuration (i.e., the relative position of various facial features), and face recognition requires fine-grained discrimination of subtle differences in facial features and their spatial configuration (Mondloch et al. 2002). Many researchers suggest that we process faces in a distinctly different way to most other objects, by integrating information across the whole face rather than processing isolated parts in a serial manner (see Maurer et al. 2002). As such, it is possible that the effects of oxytocin are not related to the social nature of faces, but rather the "special" perceptual processing that they undergo. Faces differ from houses, scenes, and objects in other ways, too – for example, faces carry additional information about emotion, attention, and membership of social groups (e.g., age, gender, ethnicity). Very few studies have directly compared the effects of oxytocin across multiple stimulus categories, so it is difficult to say whether stronger effects might be found for other categories of visual stimulus (e.g., living things such as animals or human bodies) or other stimuli that carry strong emotional, attentional, or group membership cues (e.g., non-social scenes with strong positive or negative connotations, images that direct attention in a particular direction, or images that contain non-facial cues to group membership).

WHY DOES OXYTOCIN IMPROVE MEMORY FOR FACES?

Several hypotheses have been proposed to account for the effects of oxytocin on human social cognition. On the one hand, it is possible that oxytocin improves facial expression recognition and facial identity memory by acting directly on the brain regions responsible for face-processing. Haxby, Hoffman, and Gobbini (2000) proposed a neural model of face-processing involving an interconnected network of brain regions, each responsible for a different aspect of face-processing (see also Ishai 2008). Haxby et al. (2000) distinguished between a "core" network responsible for the visual analysis of faces, and an "extended" system that connects with other neural systems (e.g., those responsible for emotional processing or recall of semantic information about individuals). Within the core system, the lateral fusiform gyrus is thought to be integral to the perception of unique identity (i.e., extracting information used to make identity judgements across changes in expression, lighting, and other superficial changes to facial appearance); whereas the superior temporal sulcus is involved in processing "changeable" aspects of a face such as expression and eye gaze. Both the lateral fusiform gyrus and the superior temporal sulcus are connected bilaterally to the amygdala, an area of the extended face-processing network that is strongly implicated in emotion processing (particularly fear processing) in humans (see Phelps 2006).

Studies using functional imaging techniques support the idea that intranasal inhalation of oxytocin can modulate activity in both core and extended regions of the face-processing network. Currently no fMRI studies have specifically examined the effects of oxytocin during face recognition tasks, but some studies have used faces as stimuli during free-viewing and emotion processing tasks. There is some evidence of effects of oxytocin in core regions of the face-processing system: Domes et al. (2007b) found oxytocin-induced modulation of activity in the superior temporal lobe in males, and Domes et al. (2010) found wide-ranging effect of oxytocin in the fusiform and superior temporal gyrus in females. In terms of extended regions, a large number of studies have found that oxytocin modulates activity in the amygdala (see Bethlehem, van Honk, Auyeung and Baron-Cohen 2013, Kanat, Heinrichs and Domes 2014, for reviews), potentially accounting for some of the effects of oxytocin on emotion processing and fluctuation of face memory results across emotions. Similar to behavioural results, though, there is significant variability in fMRI studies, particularly in terms of gender (Domes et al. 2007b, Domes et al. 2010) and emotional valence (Domes et al. 2007b, Gamer et al. 2010). Alongside extensive behavioural evidence that oxytocin acts on a wide variety of social cognitive processes beyond face-processing (see Bartz et al. 2011, Evans et al. 2014, Guastella and MacLeod 2012, for reviews), these findings suggest that the underlying mechanism by which oxytocin affects face recognition may be more complex than simple modulation of activity within the face-processing system.

Indeed, several authors have criticised the idea that the effects of oxytocin are specific to a number of discrete, higher-order cognitive processes, such as social judgements or memory for social stimuli. Instead, it has been suggested that oxytocin acts on broader affective and cognitive mechanisms, by increasing the salience of social stimuli (Bartz et al. 2011, Shamay-Tsoory et al. 2009), modulating approach-withdrawal motivation (that is, increasing behaviours associated with social approach and decreasing those associated with social withdrawal; Kemp and Guastella 2011), or reducing anxiety (specifically social anxiety, Churchland and Winkielman 2012). These hypotheses propose that oxytocin induces general changes that can manifest themselves as specific behavioural effects, such as increased sensitivity or accuracy to facial expressions of emotion or enhanced memory for faces (or, in other social-cognitive domains, changes in social evaluations of in-group and out-group members). For example, the social salience hypothesis suggests that enhancing the salience of social stimuli (e.g., faces) will selectively improve processing of social cues in the environment; similarly, the anxiolytic hypothesis proposes that reducing anxiety levels will make individuals more willing to engage in social interactions and look people directly in the eyes. Each of these hypotheses has a significant amount of behavioural and neural support, but currently it is difficult to discriminate between them: for example, the modulatory effects of oxytocin on visual attention and amygdala activity (e.g., Domes et al. 2007b, Gamer et al. 2010, Guastella et al. 2008b) are consistent with the social salience and anxiolytic hypotheses; differing effects of emotional valence (in both expression processing and face recognition tasks) are consistent with the anxiolytic and approach-withdrawal hypotheses (see Bartz et al. 2011, Kemp and Guastella 2011); and individual differences in response to oxytocin (e.g., Bartz et al. 2010, Bate et al. 2014, Domes et al. 2007a) can be accounted for by all three hypotheses. Further, these are not exclusive hypotheses: oxytocin may have widespread effects that encompass anxiety reduction and modulation of social salience or approach-withdrawal behaviours. As such, it remains unclear what underlying mechanisms are responsible for the effect of oxytocin on face memory.

CONCLUSION

In conclusion, while there is some evidence that oxytocin can improve face recognition abilities, there is still a large gap in our knowledge about how and why these effects emerge. Currently, the greatest challenge in this field is developing a greater understanding of the factors that moderate the effects of oxytocin, both within individuals (e.g., different types of stimuli or task demands) and between individuals (e.g., gender and baseline face recognition ability). On a practical level, further research is needed using carefully designed experiments to differentiate between face perception, encoding, and consolidation; and to compare effects across multiple stimulus categories and groups of individuals, so we can fully understand the locus and breadth of the effects of oxytocin on face-processing. On a more theoretical level, it may be useful to examine approach and withdrawal behaviours (e.g., inducing compared to perceiving anger, Kemp and Guastella 2011); and incorporate more fine-grained or sensitive measures of anxiety, particularly social anxiety, than the questionnaires generally used in oxytocin research (Churchland and Winkielman 2012).

This will allow researchers to develop a clearer understanding of the potential for oxytocin to be used as a therapeutic intervention for individuals who experience face recognition difficulties.

Reviewed by Dr Benjamin Parris, Bournemouth University, UK
(bparris@bournemouth.ac.uk).

REFERENCES

Bate, S., Bennetts., R. J., Bindemann, M., Udale, R., and Bussant, A. (2015). Oxytocin increases bias, but not accuracy, in face recognition line-ups. *Social Cognitive and Affective Neuroscience*, 10, 1010-1014.

Bate, S., Cook, S. J., Duchaine, B., Tree, J. J., Burns, E. J., and Hodgson, T. L. (2014). Intranasal inhalation of oxytocin improves face processing in developmental prosopagnosia. *Cortex*, 50, 55-63.

Bartz, J. A., Zaki, J., Bolger, N., Hollander, E., Ludwig, N. N., Kolevzon, A., and Ochsner, K. N. (2010). Oxytocin selectively improves empathic accuracy. *Psychological Science*, 21, 1426-1428.

Blandon-Gitlin, I., Pezdek, K., Saldivar, S., and Steelman, E. (2014). Oxytocin eliminates the own-race bias in face recognition memory. *Brain research*, 1580, 180-187.

Bowles, D. C., McKone, E., Dawel, A., Duchaine, B. C., Palermo, R., Schmalzl, L., et al. (2009). Diagnosing prosopagnosia: Effects of ageing, sex, and participant-stimulus ethnic match on the Cambridge Face Memory Test and Cambridge Face Perception Test. *Cognitive Neuropsychology*, 26, 423-455.

Bruce, V., and Young, A. (1986). Understanding face recognition. *British Journal of Psychology*, 77, 305–327.

Churchland, P. S., and Winkielman, P. (2012). Modulating social behavior with oxytocin: how does it work? What does it mean? *Hormones and behavior*, 61(3), 392-399.

D'Argembeau, A., Van der Linden, M., Comblain, C., and Etienne, A. M. (2003). The effects of happy and angry expressions on identity and expression memory for unfamiliar faces. *Cognition and Emotion*, 17(4), 609-622.

Di Simplicio, M., Massey-Chase, R., Cowen, P. J., and Harmer, C. J., (2009). Oxytocin enhances processing of positive versus negative emotional information in healthy male volunteers. *Journal of Psychopharmacology*, 23, 241–248.

Domes, G., Heinrichs, M., Michel, A., Berger, C., and Herpertz, S. C. (2007a). Oxytocin improves "mind-reading" in humans. *Biological psychiatry*, 61(6), 731-733.

Domes, G., Heinrichs, M., Gläscher, J., Büchel, C., Braus, D. F., and Herpertz, S. C. (2007b). Oxytocin attenuates amygdala responses to emotional faces regardless of valence. *Biological Psychiatry*, 62(10), 1187-1190.

Domes, G., Lischke, A., Berger, C., Grossmann, A., Hauenstein, K., Heinrichs, M., and Herpertz, S. C. (2010). Effects of intranasal oxytocin on emotional face processing in women. *Psychoneuroendocrinology*, 35, 83-93.

Duchaine, B., and Nakayama, K. (2006). The Cambridge Face Memory Test: Results for neurologically intact individuals and an investigation of its validity using inverted face stimuli and prosopagnosic participants. *Neuropsychologia*, 44, 576-585.

Evans, S. L., Dal Monte, O., Noble, P., and Averbeck, B. B. (2014). Intranasal oxytocin effects on social cognition: a critique. *Brain research*, 1580, 69-77.

Fehm-Wolfsdorf, G., and Born, J. (1991). Behavioral effects of neurohypophyseal peptides in healthy volunteers: 10 years of research. *Peptides*, 12(6), 1399-1406.

Ferguson, J. N., Young, L. J., and Insel, T. R. (2002). The neuroendocrine basis of social recognition. *Frontiers in neuroendocrinology*, 23(2), 200-224.

Gamer, M., Zurowski, B., and Büchel, C. (2010). Different amygdala subregions mediate valence-related and attentional effects of oxytocin in humans. *Proceedings of the National Academy of Sciences of the USA*, 107, 9400-9405.

Guastella, A. J., Mitchell, P. B., and Mathews, F. (2008a). Oxytocin enhances the encoding of positive social memories in humans. *Biological psychiatry*, 64(3), 256-258.

Guastella, A. J., Mitchell, P. B., and Dadds, M. R. (2008b). Oxytocin increases gaze to the eye region of human faces. *Biological Psychiatry*, 63, 3-5.

Guastella, A. J.,MacLeod, C. (2012).A critical review of the influence of oxytocin nasal spray on social cognition in humans: evidence and future directions. *Hormones and Behavior*, 61, 410–418.

Haxby, J. V., Hoffman, E. A., and Gobbini, M. I. (2000). The distributed human neural system for face perception. *Trends in Cognitive Science*, 4, 223-233.

Heinrichs, M., Meinlschmidt, G., Wippich, W., Ehlert, U., and Hellhammer, D. H. (2004). Selective amnesic effects of oxytocin on human memory. *Physiology and Behavior*, 83(1), 31-38.

Heinrichs, M., von Dawans, B., and Domes, G. (2009). Oxytocin, vasopressin, and human social behavior. *Frontiers in Neuroendocrinology*, 30, 548-557.

Herzmann, G., Bird, C. W., Freeman, M., and Curran, T. (2013). Effects of oxytocin on behavioral and ERP measures of recognition memory for own-race and other-race faces in women and men. *Psychoneuroendocrinology*, 38(10), 2140-2151.

Herzmann, G., Young, B., Bird, C. W., and Curran, T. (2012). Oxytocin can impair memory for social and non-social visual objects: a within-subject investigation of oxytocin's effects on human memory. *Brain Research*, 1451, 65-73.

Ishai, A. (2008). Let's face it: it's a cortical network. *Neuroimage*, 40(2), 415-419.

Jacoby, L. L. (1991). A process dissociation framework: Separating automatic from intentional uses of memory. *Journal of Memory and Language*, 30, 513-541.

Kanat, M., Heinrichs, M., and Domes, G. (2014). Oxytocin and the social brain: Neural mechanisms and perspectives in human research. *Brain research*, 1580, 160-171.

Kemp, A. H., and Guastella, A. J. (2011). The role of oxytocin in human affect: A novel hypothesis. *Current Directions in Psychological Science*, 20, 222-231.

Lischke, A., Berger, C., Prehn, K., Heinrichs, M., Herpetz, S. C., and Domes, G. (2012). Intranasal oxytocin enhances emotion recognition from dynamic facial expressions and leaves eye-gaze unaffected. *Psychoneuroendocrinology*, 37, 475-481.

Maurer, D., Le Grand, R., and Mondloch, C. J. (2002). The many faces of configural processing. *Trends in Cognitive Sciences,* 6(6), 255–260.

Mondloch, C. J., Le Grand, R., and Maurer, D. (2002). Configural face processing develops more slowly than featural face processing. *Perception,* 31, 553–566.

Peterson, M. F., and Eckstein, M. P. (2012). Looking just below the eyes is optimal across face recognition tasks. *Proceedings of the National Academy of Sciences*, 109(48), E3314-E3323.

Phelps, E. A. (2006). Emotion and cognition: insights from studies of the human amygdala. *Annual Review of Psychology*, 57, 27-53.

Rimmele, U., Hediger, K., Heinrichs, M., and Klaver, P. (2009). Oxytocin makes a face in memory familiar. *Journal of Neuroscience*, 29, 38-42.

Savaskan, E., Ehrhardt, R., Schulz, A., Walter, M., and Schachinger, H. (2008). Post-learning intranasal oxytocin modulates human memory for facial identity. *Psychoneuroendocrinology,* 33, 368-374.

Shahrestani, S.,Kemp, A. H., Guastella, A. J. (2013). The impact of a single administration of intranasal oxytocin on the recognition of basic emotions in humans: a meta-analysis. *Neuropsychopharmacology,* 38,1929–1936.

Shamay-Tsoory, S. G., Fischer, M., Dvash, J., Harari, H., Perach-Bloom, N., and Levkovitz, Y. (2009). Intranasal administration of oxytocin increases envy and Schadenfreude (Gloating). *Biological Psychiatry,* 66, 864-870.

Skuse, D. H., Lori, A., Cubells, J. F., Lee, I., Conneely, K. N., Puura, K., ... and Young, L. J. (2014). Common polymorphism in the oxytocin receptor gene (OXTR) is associated with human social recognition skills. *Proceedings of the National Academy of Sciences,* 111(5), 1987-1992.

Striepens, N., Kendrick, K. M., Hanking, V., Landgraf, R., Wüllner, U., Maier, W., and Hurlemann, R. (2013). Elevated cerebrospinal fluid and blood concentrations of oxytocin following its intranasal administration in humans. *Scientific Reports,* 3, 3440.

Susilo, T., and Duchaine, B. (2013). Advances in developmental prosopagnosia research. *Current Opinion in Neurobiology,* 23, 423-429.

Van IJzendoorn, M. H., and Bakermans-Kranenburg, M. J. (2012). A sniff of trust: meta-analysis of the effects of intranasal oxytocin administration on face recognition, trust to in-group, and trust to out-group. *Psychoneuro-endocrinology,* 37(3), 438-443.

Weigelt, S., Koldewyn, K., and Kanwisher, N. (2012). Face identity recognition in autism spectrum disorders: a review of behavioral studies. *Neuroscience and Biobehavioral Reviews,* 36(3), 1060-1084.

Weisman, O., and Feldman, R. (2013). Oxytocin effects on the human brain: Findings, questions, and future directions. *Biological Psychiatry,* 74, 158-159.

Wilmer, J. B., Germine, L., Chabris, C. F., Chatterjee, G., Gerbasi, M., and Nakayama, K. (2012). Capturing specific abilities as a window into human individuality: the example of face recognition. *Cognitive neuropsychology,* 29(5-6), 360-392.

Young, S. G., Hugenberg, K., Bernstein, M. J., and Sacco, D. F. (2012). Perception and motivation in face recognition: a critical review of theories of the cross-race effect. *Personality and Social Psychology Review,* 16, 116-142.

Zaki, J., Bolger, N., and Ochsner, K. N. (2011). Social effects of oxytocin in humans: Context and person matter. *Trends in Cognitive Sciences,* 15, 301-309.

In: Advances in Oxytocin Research
Editor: Evan Anderson

ISBN: 978-1-63483-991-4
© 2015 Nova Science Publishers, Inc.

Chapter 4

ASSOCIATION AMONG THE AGING PROCESS, OBESITY, SEXUAL AND FEEDING BEHAVIOR AND OXYTOCIN LEVELS: AN INTERFACE WITH CORTISOL

Laura Schlatter Rosemberg[1,*]
Camila Bittencourt Jacondino[2,†]
and Maria Gabriela Valle Gottlieb[3,‡]

[1]Nutritionist. PhD student Biomedical Gerontology by the Pontifical Catholic University of Rio Grande do Sul, Brazil
[2]Nurse. PhD student Biomedical Gerontology by the Pontifical Catholic University of Rio Grande do Sul, Brazil
[3]Biogerontologist. Researcher and professor of the Institute of Geriatrics and Gerontology (IGG), PUCRS, Porto Alegre, Brazil

ABSTRACT

The increase in the elderly population has drawn attention to diseases whose incidence is high in this age group, such as obesity. A common characteristic linked to the aging process is the modification of body

---

[*] Email: laurarosemberg@hotmail.com.
[†] Email: camilabjacondino@gmail.com.
[‡] Email: maria.gottlieb@pucrs.br.

composition, mainly increased fat mass. Age-related obesity has a multifactorial origin, mainly involving the interaction of genetic, hormonal and environmental factors. Recent literature has suggested that the pathophysiology of obesity is related to not only neuroendocrine disorders involving the control of appetite and satiety but also neuroendocrine routes that are common to psychiatric (depression and anxiety) and behavioral disorders. In this sense, the hormone oxytocin can be the trigger for the behavioral changes associated with aging, sexual and feeding behavior, as these behaviors share common reward mechanisms involved in the stimulation and feeling of pleasure. Therefore, it is important to emphasize that with aging, disorders may occur in the synthesis and secretion of oxytocin, which may play an etiologic role in not only sexual behavior but also in food and energy metabolism, thus favoring the onset of obesity. Oxytocin is synthesized in the paraventricular and supraoptic nuclei of the hypothalamus that acts in the hippocampus and the amygdala. Therefore, oxytocin signaling can influence the release of other hormones such as cortisol. Because cortisol also acts in the hippocampus and the amygdala, these two hormones can interact. However, information about this correlation between oxytocin and cortisol with respect to sexual and eating behaviors and obesity in the elderly are scarce and controversial. Oxytocin seems to be a component of the intricate network of neurophysiological processes underlying sexual responses. The literature has shown that oxytocin plays various roles in sexual responses, such as uterine contraction, ejaculation, orgasm and feeling of pleasure. In this context, one cannot rule out that dietary intake can also provide a feeling of pleasure and an anti-stress effect similar to the sexual act and that cortisol and oxytocin exert effects on the central nervous system. Therefore, unifying the information regarding the functions performed by oxytocin, it seems plausible that there is an association among aging, obesity (via feeding behavior), and sexual behavior and oxytocin levels. It is within this context that this manuscript aims to contribute to a better understanding of the mechanisms involved in the age-related changes in body composition, sexual, and feeding behavior, in which oxytocin and cortisol probably play a relevant role.

Keywords: aging, oxytocin, cortisol, sexual behavior, eating behavior, Obesity

1. INTRODUCTION

Aging is an inevitable and inexorable process that involves several changes, from the molecular to the morphophysiological level [1]. One of the most common findings associated with the aging process is the modification of body composition. This alteration, especially when related to reduced skeletal muscle mass and increased body fat, is an inherent characteristic of the aging process [2].

Therefore, it is expected that the elderly experience a reduction in lean body mass and an increase in fat mass [3]. Due to the loss of muscle mass, the basal metabolic rate decreases by 4% per decade from 50 years of age [4]. Another process associated with aging is the redistribution of adipose tissue, including the accumulation of abdominal and visceral fat [5]. Even when the body weight is stable, people tend to gain weight with age because the decrease in muscle mass is replaced by fat mass. In addition, the remaining muscle can be infiltrated by fat [6].

The increase in the elderly population has drawn attention to diseases whose incidence is high in this age group, such as obesity [1]. Research between 2007 and 2010 in the United States showed that the prevalence of obesity among Americans aged 65 years or more was approximately 35%. There are more than 8 million obese people aged between 65 and 74 years and nearly 5 million obese people aged 75 years or more [7]. In contrast to some developed countries, Brazil does not systematically perform national population surveys. Data from the Survey of Family Budgets (POF) of 2002-2003 showed that 8.9% of Brazilian men and 13.1% of Brazilian women are obese. These values vary widely according to the region and the economic status of each area of the country [8].

Obesity has a multifactorial origin, mainly involving the interaction of genetic and environmental factors [9]. Recent literature has suggested that the pathophysiology of obesity is related to not only neuroendocrine disorders that involve control of appetite, satiety and overeating but also neuroendocrine routes that are common to psychiatric (depression and anxiety) and behavioral disorders. In this sense, the hormones oxytocin and cortisol seem to be triggers for the behavioral changes related to feeding behavior, obesity and sexual behavior [10, 11].

Changes in sexual behavior very frequently arise during the aging process. The biological barriers often imposed by aging, such as changes in physical appearance, erectile dysfunction, lack of libido, vaginal dryness, and chronic diseases (e.g., obesity), can be influenced directly or indirectly by hormonal pathways.

Therefore, it is important to highlight that with aging, there might be changes in the synthesis and secretion of cortisol and oxytocin, which are described in the literature to be involved in not only sexual and eating habits but also chronic diseases, such as obesity [12].

The literature on cortisol is wider and indicates that the levels of this hormone are increased throughout the aging process. Furthermore, cortisol induces increased glucose and fatty acid levels in the blood circulation as well as increased blood pressure, which may be associated with obesity [13]. Oxytocin acts as a natriuretic agent by participating in osmolar adjustment, increasing sodium excretion by the kidneys, providing glomerular filtration and, thus, affecting the blood pressure levels [14].

Parallel studies suggest that sexual stimulation increases the cortisol levels in men to direct their energy toward a sexual situation [15]. Furthermore, in other studies, cortisol inhibited sexual responses to mechanical stress; this finding suggests a role of cortisol antagonism in sexual behavior [16, 17, 18].

Because cortisol acts in the hippocampus and the amygdala, it can be influenced by or act antagonistically or synergistically with oxytocin, which is also present in these regions.

However, data regarding the correlation between oxytocin and cortisol in sexual behavior and obesity remain scarce and controversial. The available evidence indicates that the amygdala plays various roles, particularly in eating and sexual behavior [19, 20].

Oxytocin seems to be another component of the intricate network of neurophysiological processes of sexual responses. The literature has shown that oxytocin plays several roles related to sexual responses, such as uterine contraction, ejaculation, orgasm and sensation of pleasure [21].

In this context, we cannot disregard that food intake can also provide a feeling of pleasure and can exert anti-stress effect similar to the sexual act and to the effects of cortisol and oxytocin in the central nervous system. Searching for an isolated action of oxytocin that elicits a certain type of emotion or feeling seems to be a fruitless path [22].

Considering the putative roles played by cortisol and oxytocin, it seems plausible that there is an association between obesity (via feeding behavior), sexual behavior and the levels of these hormones in the elderly. However, the influence of oxytocin and cortisol on sexual and eating behavior (obesity) that may be associated with aging remains a part of a large puzzle that has yet to be assembled.

The present article intends to contribute to a better understanding of these interactions. Figure I show a hypothetical diagram of these interactions.

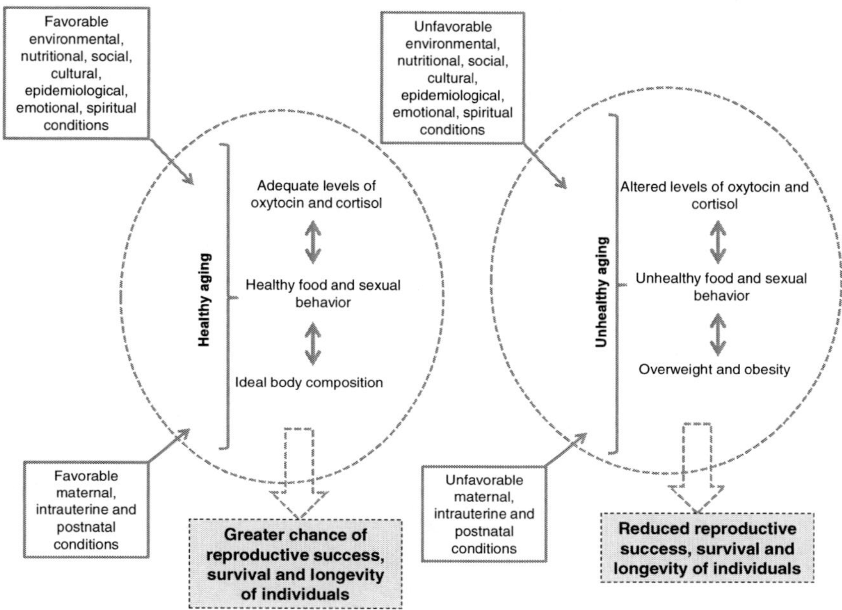

Figure 1. Hypothetical diagram of interactions and effect of the extrinsic factors on oxytocin, cortisol, feeding and sexual behavior associated with obesity and its impact on human reproduction, survival and longevity.

2. BIOLOGICAL ASPECTS OF AGING: AN APPROACH TO SEXUAL BEHAVIOR AND OBESITY

The term aging is a difficult concept. However, it is often used to describe the morphological and functional changes that occur after sexual maturation and that progressively undertake the responsiveness of individuals to environmental stress and the maintenance of homeostasis [23]. Finch [24] noted that aging is the only universal feature associated with these changes over time, regardless of whether these changes have a deleterious effect on vitality and longevity. The term aging is also used to describe virtually all time-dependent changes to which biological entities, from molecules to ecosystems, are subject, although the mechanisms of aging and their consequences on function can be quite different [25].

In demographic terms, the consistently accelerating growth of the elderly segment of the population is a phenomenon that affects most developed and developing countries. In the case of Brazil, the process of population aging began to occur in the mid-20th century, when there was a reversal of the fertility and mortality curves, thus increasing the life expectancy of the Brazilian population. This increase in life expectancy can result in the emergence of so-called chronic diseases (CD), most of which are multifactorial diseases, which involve the interaction of several biological, psychosocial and environmental variables. Importantly, these diseases began to become increasingly prevalent in the elderly population, and this increase has caused a great burden to the State. These diseases can lead to the accumulation of damage due to prolonged exposure to risk factors, long life, and unfavorable genetic-metabolic conditions. Notably, the aging process itself causes changes at different morphophysiological levels, highlighting the changes in energy metabolism, body composition, food standards and sexual behavior in individuals. The aging process induces several physical changes in both men and in women and occasionally affects sexual ability. Regarding aging, biological factors can exert control over sexual dimension, affecting sexual desire, sexual functioning and, indirectly, sexual satisfaction [26].

During the aging process, individuals exhibit hormonal changes, mainly including a reduction in the levels of the hormone testosterone in men and progesterone in women. The levels of the hormones cortisol and oxytocin also undergo changes during this process, and these changes appear to be involved in various pathological mechanisms beyond sexual behavior, such as fat mass accumulation and lean body mass reduction, increased glucose and insulin resistance, insomnia, stress, and food intake and preferences [27, 28, 29, 30, 31, 32].

Studies on the impact of aging on oxytocin are sparse and inconclusive. The few published studies show that aging can change the transmission of oxytocin in the central nervous system but apparently not in the peripheral nervous system (the neurohypophysis). However, most of these studies were not performed in humans, but rather in experimental models [33].

Changes in body composition are common findings in the elderly because aging itself induces skeletal muscle loss and increased fat mass. Currently, there is a polarization of the nutritional status among the elderly population. In one hand, there is a loss of lean muscle mass and functionality, and on the other hand, there is an excessive increase in body fat. This increase in body fat is associated with a loss of functionality, which we refer to as obesity. Thus, obesity is a morbidity that confers increased risk to the health of the elderly and that involves the interaction between body composition, energy supply, genetic factors, hormonal metabolism and lifestyle (a sedentary lifestyle) [1. 34].

Obesity is a chronic disease that is characterized by the excessive accumulation of body fat. According to the World Health Organization, globally, there are more than 1 billion overweight adults, at least 300 million of whom are obese. At present, obesity is considered as an epidemic disease, representing a serious health problem in both developed and developing countries, including Brazil. Among the main causes of obesity are consumption of foods with high energy density or high levels of saturated fats and sugar and reduced physical activity (a sedentary lifestyle). The influence of obesity on health can be assessed by the metabolic complications associated with this disorder; dyslipidemia, type 2 diabetes and atherosclerosis are important metabolic complications related to obesity because they are risk factors for cardiovascular disease [35, 36, 37, 38].

Epidemiological studies have shown a direct positive correlation between body mass index (BMI, determined by calculating the body weight divided by height squared), risk of medical complications and mortality rate. Thus, men and women who have a BMI of > 30 are considered obese and are generally at higher risk for pathological events than those who are considered overweight or healthy (BMI between 18 and 25). Additional studies showed that the distribution of body fat, especially excess abdominal fat, is associated with increased risk of cardiovascular disease. However, precise measurements of abdominal fat require the use of imaging techniques that are costly for use in population surveys [35, 36, 37].

The body dimensions reflect the overall health and well-being of individuals and, via proper assessment of nutritional status, can be used to determine important factors for the development of chronic diseases among the elderly. The evaluation of the elderly must take into account the numerous bodily changes that this group experiencing; thus, this is a very complex evaluation [39].

Additionally, it is important to take into consideration that the mechanisms and the genetic and physiological basis of obesity remain largely unexplained. What is known is that obesity is caused by not only a deficient control of satiety, excessive eating, and a sedentary lifestyle but also by psychological or behavioral disorders. Recent studies have shown that obesity or inordinately high fat accumulation is accompanied by widespread inflammation (increased plasma levels of cytokines, such as CRP, TNF-alpha, and interleukin-6, due to the disruption of the production of these cytokines by monocytes, lymphocytes, etc.), deregulation of hormones that control weight, appetite and satiety, increased oxidative stress and genetic mutations or polymorphisms [40]. Regarding the hormones involved in controlling body weight, we can highlight leptin, ghrelin, neuropeptide Y, oxytocin and cortisol.

3. OXYTOCIN

Oxytocin is a nonapeptide hormone widely known for its roles in childbirth and lactation. The uterine contraction properties of oxytocin were described in 1906 and, until fifty years later when its sequence was determined, the research on oxytocin was focused on its roles in reproduction and lactation. In recent decades, the researchers began to investigate other possible functions of oxytocin, such as in energy metabolism and in eating and sexual behavior [41].

Oxytocin is synthesized in the hypothalamus and is released by the pituitary gland, which is a small gland measuring about 1 cm in diameter and weighing from 0.5 to 1 g. The pituitary gland is located in the brain and is based in the sella turcica. This gland is connected to the hypothalamus via the hypophyseal stalk and is divided into the following sections [42, 43]:

- *Anterior pituitary (adenopituitary):* Responsible for the secretion of growth hormone, adrenocorticotrophic hormone, prolactin, follicle-stimulating hormone, thyrotropin-releasing hormone (TSH), and luteinizing hormone;
- *Posterior pituitary lobe:* Responsible for the release of antidiuretic hormone, vasopressin and oxytocin.

The hypothalamus, which is the portion of the diencephalon that provides the walls of the third ventricle, is located below the hypothalamic sulcus, which separates the hypothalamus from the thalamus. The hypothalamus is composed more than 40 nuclei and performs a number of functions, such as the following.

- Salt control
- Reproduction
- Hormonal balance
- Biological rhythms
- Control of food intake

Oxytocin is synthesized in the pericardium, more precisely, in the cells that constitute the supraoptic and paraventricular nuclei of the central nervous system. Oxytocin is initially produced as a pro-hormone that is transported toward the nerve endings located in the neural lobe, where it remains stored until an action potential that is generated in the cell body in response to specific stimuli, thereby triggering the release of oxytocin [44, 45].

The oxytocin receptor is found in different tissues [44].

- Female reproductive system: uterus and ovaries; assists in labor
- Male reproductive system: testicles and prostate; primarily functions in ejaculation and penile erection
- Mammary gland: aids in the excretion of breast milk
- Kidney: acts as a natriuretic agent to participate in osmolar adjustment, increase sodium excretion by the kidneys, and induce glomerular filtration
- Cardiovascular system: high concentrations in the right atrium; helps in the secretion of natriuretic peptide hormone, whose main effects include reducing blood pressure and blood volume due to an increased glomerular filtration rate and consequently decreasing the resorption of sodium, thus regulating blood pressure
  Pancreas: stimulates the secretion of glucagon

The neurohormone oxytocin is released in the neurohypophysis by projections from the magnocellular circuit, whereas the neurotransmitter oxytocin is released in the synaptic cleft of parvocellular neurons, which project their axons to various areas of the brain, primarily in the brainstem. Oxytocin can remain for up to 20 minutes in the interstitium. The limbic system, which is regarded as the center of emotions, is the primary site of expression of oxytocin receptors, especially in the hippocampus and the amygdala [46]. Therefore, oxytocin has been linked to the modulation of maternal behavior, social behavior, memory and empathy (promotion of trust) and has been related to psychiatric disorders such as depression, autism, and stress disorders (anxiety and post-traumatic stress disorders) [45, 47].

One of the major stimuli eliciting the release of oxytocin is physical contact, such as hugging or sexual intercourse. The oxytocin levels reach high concentrations during and after intercourse; during ejaculation, the oxytocin concentrations increases by approximately 5-fold and returns to normal 30 minutes after the sexual act. Moreover, during breastfeeding, there is an increase in the serum oxytocin levels, and this increase is associated with the affection and bonding between mother and baby, resulting in a beneficial effect on the child's psychological development. However, oxytocin may be expressed at lower levels in patients diagnosed with major depression, schizophrenia and autism [46, 48, 49].

3.1. The Role of Oxytocin in Obesity, Food Consumption and Sexual Behavior

The paraventricular nucleus of the hypothalamus is an important brain region for the regulation of both feeding and energy expenditure, and studies have suggested that oxytocin may play a role in energy metabolism [50]. Within the paraventricular nucleus, there are different subsets of peptidergic neurons, including oxytocin, vasopressin, thyrotropin-releasing hormone and corticotrophin; these neurons send signals throughout the brain. These signals form the structural basis by which the paraventricular nucleus regulates many physiological functions, including energy homeostasis [51].

In this context, several studies suggest a role of oxytocin as a regulator of body weight. Studies have shown that a reduced number and density of oxytocinergic neurons, as well as levels of oxytocin, is associated with Prader-Willi syndrome, which causes severe hyperphagia [52, 53]. Moreover, studies have suggested that oxytocin partially mediate the anorexigenic activity of leucine [54]. Other studies show that oxytocin administration reduces food intake but that the administration of oxytocin receptor antagonists results in hyperphagia [55]. Experiments conducted in mice show that the long-term infusion of oxytocin into the central nervous system also reduces weight gain, even when on a high-fat diet. However, with respect to the acute effects of oxytocin infusion, there was no change in total food intake or the food intake pattern of these mice; instead, oxytocin infusion appeared to stimulate lipid metabolism in adipose tissue [56].

Additionally, studies have shown that male mice deficient in oxytocin receptor exhibit an obesity phenotype in adulthood that is independent of food intake and motor activity [57].

Studies using male and female oxytocin gene knockout animals showed increased weight and fat stores [58]. However, in humans, this type of experimental approach has never been reported due to ethical issues and the key role of oxytocin in reproduction, survival and perpetuation of the species. However, partial deficiency of this hormone or its receptor has been associated with the development of obesity in humans. Recent findings in humans harboring rare genetic mutations linked to monogenic obesity suggest that oxytocin may also play a role in regulating appetite in humans [59]. More recently, a study comparing obese and normal individuals with and without a diagnosis of Type 2 diabetes demonstrated that the oxytocin concentrations were significantly reduced in obese diabetic patients compared to the control subjects. In addition, individuals who displayed higher concentrations of oxytocin exhibited a lower body mass index and waist circumference [60].

Furthermore, oxytocin plays an important role in sexual behavior among primates, including humans [61, 62]. This hormone is released during orgasm in both males and females, and its levels positive correlate with the degree of contraction of the pelvis in women during orgasm. Case studies have reported that oxytocin facilitates sexual parameters, such as sharp arousal in women, orgasm in men with anorgasmia, and sexual function in men with anxiety disorders [63]. Recently, using a laboratory model, it was demonstrated that well-controlled intranasal oxytocin administration triggers the release of catecholamines into plasma, which resulted in a change in the perception of arousal during masturbation in healthy men [64]. However, this study was limited to males, and the role of oxytocin in women was not elucidated. There is evidence that oxytocin may have a differential impact on sexual behavior between sexes. From clinical experience in sexual medicine, one can speculate that males are more likely to recognize changes in performance level (erectile function, orgasmic function, and sexual satisfaction), while many women put more emphasis on their perception of social aspects of the relationship, in which oxytocin also seems to play an important role, as regards the formation of affection and bonding [65].

Therefore, it is important to take into consideration that in the context of eating disorders, binge eating stands out as a compensatory action displayed by obese individuals facing emotional and relational difficulties [66]. In such cases, studies are beginning to elucidate the role of oxytocin in the regulation of food intake, preferably intake of macronutrients and power associated with the reward system [67]. The literature suggests a relationship between eating disorders and sexuality [68], in which these individuals may replace sexual pleasure for the pleasure obtained by food intake [69]. Additionally, it is important to highlight that anxiety, depression and stress appear to be associated with obesity [70, 71, 72, 73], insecure attachment behaviors [74, 75] and lower levels of sexual satisfaction [76]. In turn, linking these behaviors seems to influence feeding behavior [77, 78, 79, 80], compensating by performing anxious and avoidant behaviors in response to low sexual satisfaction [81]. Sexual satisfaction, an important component of interpersonal sexual behavior [82], is influenced by biological and psychological aspects [83]. Several authors emphasize sexual and relational difficulties, as well as sexual dissatisfaction or lack of libido, in obese individuals [84, 85, 86]. However, due to the broad physiological role of oxytocin in the body, oxytocin may be the link between obesity and the observed alterations in human sexual behavior. However, studies on this issue remain scarce and controversial, requiring further research and understanding of the mechanisms involved in this association.

4. CORTISOL, OBESITY, FOOD CONSUMPTION AND SEXUAL BEHAVIOR

Cortisol is a glucocorticoid that is synthesized in cells of the cortex of the adrenal gland, and its production is controlled by adrenocorticotropic hormone (ACTH), which is synthesized in the anterior pituitary. ACTH is released upon stimulation by corticotrophin-releasing hormone (CRH), which is produced in the paraventricular nucleus of the hypothalamus. Cortisol release significantly increases under stress conditions. Approximately 75 to 80% of plasma cortisol is bound to carrier proteins such as transcortin, 15% is bound to albumin, and the remaining small fraction circulates as a biologically active hormone [87, 88].

Cortisol performs many functions essential to the functioning of the body, including anti-inflammatory properties. Cortisol may reduce inflammation parameters, such as the erythrocyte sedimentation rate (ESR) and the C-reactive protein levels, and the release of cytokines by immune cells. After binding to selective receptors, cortisol induces the transcription of the inhibitory protein IκBα, keeping nuclear factor-κB (NFκB) in its inactive form in the cytoplasm, which inhibits NFκB from migrating to the nucleus, where it can bind to the appropriate response elements in DNA to activate the production and secretion of cytokines; thus, cortisol contributes to immunosuppression [89, 90].

The daily rhythm of cortisol secretion is relatively stable in healthy people, with the highest concentration in the morning and a decrease throughout the day, more precisely for 16 hr. Stressful situations (psychological, social or physical) can deregulate the hypothalamic-pituitary-adrenal (HPA) axis, causing changes in the circulating cortisol levels [91].

These changes in cortisol are also observed in aging and obese individuals [92].

Cortisol secretion is regulated by the hypothalamic centers that receive signals stimulating the central nervous system and is modified by adrenergic dopaminergic and serotoninergic systems in a complicated network that is only partially understood. There is also some evidence that leptin can participate in regulating the permissive action of the adrenal axis [93].

These regulatory events result in a characteristic diurnal pattern of cortisol secretion, with high activity in the early morning hours and low activity in the afternoon; this pattern that is often disturbed in obese individuals, although there is controversy regarding this finding [94, 95].

Studies have shown that the cortisol levels are increased among obese individuals, particularly among individuals with visceral obesity. Furthermore, the increase in the cortisol levels after a mixed meal was higher in obese individuals [96].

Epidemiological studies show that the cortisol levels may be associated with abdominal obesity, stressful daily events, and excess food intake [13]. Studies in animals and humans have demonstrated that stressful situations lead to a craving for foods high in calories, especially in women [95]. Some studies suggest that at least in experimental animals, this increase in energy intake may be mediated by the increased production of cortisol; this increased energy intake may help reduce the impact of stress via a reward mechanism but may contribute to the development and maintenance of obesity [96]. A study showed that individuals with increased cortisol levels due to stress tended to eat more than those not experiencing stress [97].

The cortisol levels also appear to be associated with sexual behavior. A study of men showed no increase in the cortisol levels when they thought about sex; however, the authors found that cortisol can facilitate sexual arousal [98]. A study of elderly male monkeys found a possible association between low levels of cortisol and a decline in sexual behavior in animals [99]. However, the literature on this subject remains scarce, and few studies have attempted to establish a relationship between cortisol, obesity, nutrition and sexual behavior in older people. In 1995, a study of 81 women with pre-menopausal symptoms and 70 women with menopause determined the association of obesity with attitudes towards sexuality and the levels of several hormones, including cortisol. The results showed that in the investigated sample, 35.4% had depression, 34, 3% exhibited non-specific symptoms of depression (NSD) and 42.3% suffered from anxiety. In addition, elevated levels of cortisol have been found in women in the late stage of menopause.

Multivariate analysis showed that NSD were associated with negative attitudes towards sexuality and that the cortisol levels were associated with BMI. Another interesting finding of this study was that overweight was associated with NSD, sleep disorders and hormonal changes [100]. Thus, based on the literature, it appears that there is an association between obesity, sexual behavior and cortisol levels. However, it is not yet elucidated how obesity, eating patterns, sexual behavior, and the oxytocin and cortisol levels interact throughout the aging process.

CONCLUSION

Among the elderly, sexual and feeding behavior, the latter of which causes both obesity and malnutrition in this age group, are neglected subjects because public health policies remain too focused on the youth segment of society in terms of the development and reproduction phases. However, it is notable that the prevalence of obesity increases with age and that feeding and sexual behavior also exhibit adverse alterations with age. It is also important to mention that obesity develops due to both behavior and food preferences and that sexual behavior shares the same neuroendocrine routes involving cortisol and oxytocin that are associated with the stimulation and the sensation of pleasure. Studies on the impact of aging on oxytocin in humans are scarce. There is no evidence in the recent literature showing an association between oxytocin, cortisol, obesity, sexual behavior and feeding behavior in the elderly, most likely because of the complexity of not only these interactions but also the technical implementation, logistics and methodology appropriate for this segment of the population. That is, the information about the oxytocinergic system, cortisol, obesity, eating behavior and sexual behavior are still being produced in isolation, and it is up to us to solve this puzzle to begin to understand how these factors interact to produce a certain phenotype over the aging process.

REFERENCES

[1]   MS - Ministério da Saúde. Atenção à Saúde da Pessoa Idosa e Envelhecimento. 2010. Disponível em: <http://portal.saude.gov.br /portal/arquivos/pdf/volume12.pdf>.

[2]   Sánchez-García, S., García-Pena, C., Duque-López, M.X., Juárez-Cedillo, T., Cortes-Núnez, A.R. & Reyes-Beaman, S. (2007). Anthropometric measures and nutritional status in a healthy elderly population. BMC Public Health, 7, 1-9.

[3]   Kaur, M. & Tawar, I. (2011). Body composition and fat distribution among older Jat females: A rural-urban comparison. *HOMO*, 62, 374-85.

[4]   Mathus-Vliegen, E.M.H., Basdevant, A. & Finer, N. et al. (2012). Obesity Management Task Force of the European Association for the Study of Obesity. Prevalence, pathophysiology, health consequences and treatment options of obesity in the elderly: a guideline. *Obes Facts, 5,* 460–483.

[5]   Mastroeni, M.F., Mastroeni, S.S., Erzinger, G.S. & Marucci, M.F. (2010). Antropometria de idosos residentes no município de Joinville-SC, Brasil. *Rev. Bras Geriatr. Gerontol.,* 13, 29-40.

[6]   Cervi, A., Franceschini, S. & Priore, S. (2005). Critical analysis of the use of the body mass index for the elderly. *Rev. nutr.,* 18, 765-775.

[7]   Fakhouri, T.H., Ogden, C.L., Carroll, M.D., Kit, B.K., & Flegal, K.M. (2012). Prevalence of obesity among older adults in the United States, 2007-2010. NCHS Data Brief, 1-8.

[8]   IBGE - Instituto Brasileiro de Geografia e Estatística. (2004). Pesquisa de orçamentos familiares 2002-2003: análise da disponibilidade domiciliar de alimentos e do estado nutricional no Brasil.

[9]   Grundy, S.M. (1998). Multifactorial causation of obesity: implications for prevention. *Am. J. Clin. Nutr.,* 67, 563S–572S.

[10] Vicennati, V., Pasqui, F. & Cavazza, C. et al. (2011). Cortisol, energy intake, and food frequency in overweight/obese women. *Nutrition, 27,* 677–680.

[11] Rehman, H.U. & Masson, E.A. (2001). Neuroendocrinology of ageing. *Age and Ageing,* 30, 279–287.

[12] Ginsberg, T.B. (2006). Aging and sexuality. *The Medical clinics of North America,* 90, 1025-1036.

[13] Vicennati, V., Pasqui, F., Cavazza, C., Pagotto, U. & Pasquali R. (2009). Stress-related development of obesity and cortisol in women. Obesity (Silver Spring), 17, 1678– 1683.

[14] Petersson, M., Alster, P., Lundeberg, T. & Uvnas-Moberg, K. (1996). Oxytocin causes long-term decrease of blood pressure in female and male rats. *Physiology and Behavior,* 611, 1311-1315.

[15] Goldey, K.L. & Anders, S.M. (2012). Sexual Thoughts: Links to Testosterone and Cortisol in Men. Arch Sex Behav, 41, 1461–1470.

[16] Kirschbaum, C. & Hellhammer, D.H. (1989). Salivary cortisol in psychobiological research: An overview. *Neuropsychobiology,* 22, 150–169.

[17] Bodenmann, G., Atkins, D.C., Schar, M. & Poffet, V. (2010). The association between daily stress and sexual activity. *Journal of Family Psychology,* 24, 271–279.

[18] Bodenmann, G., Ledermann, T., Blattner, D. & Galluzzo, C. (2006). Associations among everyday stress, critical life events, and sexual problems. *Journal of Nervous and Mental Disease,* 194:494–501.

[19] Mestor, C.M. & Frohlich, P.F. (2000). The neurobiology of sexual function. *Arch. Gen. Psychiatry,* 57, 1012-1030.

[20] Salamon, E. et al. (2005). Role of amygdala in mediating sexual and emotion behavior via coupled nitric oxide release. *Acta Pharmacol. Sin.,* 26, 389-395.

[21] Carmichael, M.S. et al. (1987). Plasma oxytocin increases in the human sexual response. *J. Clin. Endocrinol. Metab.,* 64, 27-31.

[22] Dacome, O.A. & Garcia, R.F. (2008). Efeito modulador da ocitocina sobre o prazer. Revista Saúde e Pesquisa, 1, 193-200.

[23] Troen R.B. (2003). The Biology Of Aging. *The Mount Sinai Journal of Medicine,* 70, 3-22.

[24] Finch, C.E. (1990). Longevity, Senescence and the Genome. (Chicago: University of Chicago Press).

[25] Oliveira, K.A., Da Cunha, G.L., Da Cruz, I.B.M., et al. (1999). Estudo da longevidade e envelhecimento no modelo experimental Drosophila melanogaster. In: Sacchet, F.A.M., org. Genética para que te quero? (195-200). Porto Alegre: Universidade Federal do Rio Grande do Sul.

[26] Vieira, K.F.L. (2012). Sexualidade e qualidade de vida do idoso: desafios contemporâneos e repercussões psicossociais [Tese]. João Pessoa (PB): Universidade Federal da Paraíba/Universidade Federal do Rio Grande do Norte.

[27] Gibson, E.L., Checkley, S., Papadopoulos, A., Poon, L., Daley, S., & Wardle, J. (1999). Increased salivary cortisol reliably induced by a protein-rich midday meal. *Psychosom. Med.,* 61:214–224.

[28] Slag, M.F., Ahmad, M., Gannon, M.C., & Nuttall, F.Q. (1999). Meal stimulation of cortisol secretion: a protein induced effect. *Metabolism,* 30, 1104–1108.

[29] Vicennati, V., Ceroni, L., Gagliardi, L., Gambineri, A., & Pasquali R. (2002). Response of the hypothalamic-pituitary-adrenocortical axis to

high-protein/fat and highcarbohydrate meals in women with different obesity phenotypes. *J. Clin. Endocrinol. Metab.*, 87, 3984–3988.

[30] Vicennati, V. & Pasquali, R. (2000). Abnormalities of the hypothalamic-pituitary-adrenal axis in nondepressed women with abdominal obesity and relations with insulin resistance: evidence for a central and peripheral alteration. *J. Clin. Endocrinol. Metab.*, 85, 4093–4098.

[31] Maejima, Y., Iwasaki, Y., Yamahara, Y., Kodaira, M., Sedbazar, U., & Yada, T. (2011). Peripheral oxytocin treatment ameliorates obesity by reducing food intake and visceral fat mass. *Aging*, 3, 1169-1177.

[32] Gordon, I., Zagoory, S.O., Scheneiderman, I., Leckman, J.F., Weller, A., & Feldman, R. (2008). Oxytocin and cortisol in romantically unattached young adults: associations with bonding and psychological distress. *Psychophysiology*, 45, 349-352.

[33] Ebner, N.C., Maura, G.M., Macdonald, K., Westberg, L., & Fischer, H. (2013). Oxytocin and socioemotional aging: current knowledge and future trends, *Front. Hum. Neurosci.*, 7, 487.

[34] Veras, R. (2009). Envelhecimento populacional contemporâneo: demandas, desafios e inovações. *Rev. Saúde Publica*, 43, 548-554.

[35] Rexrode, K.M., Hennekens, C.H., Willett, W.C., Colditz, G.A., Stampfer, M.J., Rich-Edwards, J.W., et al. (1997). A prospective study of body mass index, weight change, and risk of stroke in women. *JAMA*, 277, 1539–1545.

[36] Oliveria, S.A., Felson, D.T., Cirillo, P.A., Reed, J.I., & Walker, A.M. (1999). Body weight, body mass index, and incident symptomatic osteoarthritis of the hand, hip, and knee. *Epidemiology*, 10, 161–166.

[37] WHO - World Health Organization. (2000). Obesity: Preventing and managing the global epidemic. Geneva.

[38] James, P.T., Leach, R., Kalamara, E., & Shayeghi, M. (2001). The worldwide obesity epidemic. *Obes Res.*, 9, 228S–233S.

[39] Barbosa, A., Souza, J., Lebrão, M., Laurenti, R., & Marucci, M. (2005). Anthropometry of elderly residents in the city of São Paulo, Brazil. *Cad Saúde Pública*, 21, 1929-1938.

[40] Corvera, S., Burkart, A., Kim, J.Y., Christianson, J., Wang, Z., & Scherer, P.E. (2006). Keystone meeting summary: 'Adipogenesis, obesity, and inflammation' and 'Diabetes mellitus and the control of cellular energy metabolism.' January 21–26, 2006, Vancouver, Canada. *Genes Dev.*, 20, 2193–2201.

[41] Lee, H., Macbeth, A., Pagani, J., & Young, W. (2009). Oxytocin: The great facilitator of life. *Prog. Neurobiol.*, 88, 127–151.

[42]   Barberis, C. & Tribollet, E. (1996). Vasopressin and oxytocin receptores in the central nervous system. *Crit. Ver. Neurobiol.,* 10, 119-154.

[43]   Fontes, R.F. (2012). Ação central da ocitocina no controle da ingestão alimentar: influência do ciclo estral [Dissertação]. Aracaju (SE): Universidade Federal de Sergipe.

[44]   Gimpl, G. & Fahrenholz, F. (2001). The oxytocin receptor system: structure, function, and regulation. Physiological Reviews, 81:629–683.

[45]   Leng, G., Meddle, S. & Douglas, A.J. (2008). Oxytocin and the maternal brain. *Currente opinion in Pharmacology,* 8, 731-734.

[46]   Matsuzaki, M., Matsushita, H., Tomizawa, K., & Matsui, H. (2012). Oxytocin: A therapeutic target for mental desorders. *J. Physiol.,* 62, 441-444.

[47]   Sanchez, T.A. & Frederica, H. (2012). Oxitocina-vasopressina: el futuro em tratamentos. *Psicofarmacol,* 12, 9-14.

[48]   Carmichael, M.S., Humbert, R., Dixen, J., Palmisano, G., Greenleaf, W., & Davidson, J.M. (1987) Plasma oxytocin increases in the human sexual response. *J. Clin. Endocrinol. Metab.,* 64:27–31.

[49]   Blaicher, W., Gruber, D., Bieglmayer, C., & Blaicher, A.M. (1999). The role of oxytocin in relation to female sexual arousal. *Gynecol. Obstet. Invest,* 47, 125-126.

[50]   Madden, C.J. & Morrison, S.F. (2009). Neurons in the paraventricular nucleus of the hypothalamus inhibit sympathetic outflow to brown adipose tissue. *Am. J. Physiol. Regul. Integr. Comp. Physiol.,* 296, R831–843.

[51]   Olszewski, P.K., Klockars, A., Schioth, H.B., & Levine, A.S. (2010). Oxytocin as feeding inhibitor: maintaining homeostasis in consummatory behavior. *Pharmacol. Biochem. Behav.,* 97, 47–54.

[52]   Martin, A., State, M., Anderson, G.M., Kaye, W.M., Hanchett, J.M., et al. (1998). Cerebrospinal fluid levels of oxytocin in Prader-Willi syndrome: a preliminar report. *Biol. Psychiatry,* 44:1349–1352.

[53]   Swaab, D.F., Purba, J.S., & Hofman, M.A. (1995). Alterations in the hypothalamic paraventricular nucleus and its oxytocin neurons (putative satiety cells) in Prader-Willi syndrome: a study of five cases. *J. Clin. Endocrinol. Metab.,* 80, 573–579.

[54]   Blouet, C., Jo, Y.H., Li, X., & Schwartz, G.J. (2009). Mediobasal hypothalamic leucine sensing regulates food intake through activation of a hypothalamus-brainstem circuit. *J. Neurosci.,* 29:8302–8311.

[55]   Zhang, G., Bai, H., Zhang, H., Dean, C., Wu, Q., et al. (2011). Neuropeptide exocytosis involving synaptotagmin-4 and oxytocin in

hypothalamic program- ming of body weight and energy balance. *Neuron,* 69:523–535.

[56] Deblon, N., Veyrat-Durebex, C., Bourgoin, L., Caillon, A., Bussier, A.L., Petrosino, S., et al. (2011). Mechanisms of the anti-obesity effects of oxytocin in diet-induced obese rats. *Plos One,* 6, e25565.

[57] Takayanagi, Y., Kasahara, Y., Onaka, T., Takahashi, N., Kawada, T., & Nishimori, K. (2008). Oxytocin receptor-deficient mice developed late-onset obesity. *Neuro report,* 19, 951–955.

[58] Nishimori, K., Young, L.J., Guo, Q., Wang, Z., Insel, T.R., & Matzuk, M.M. (1996). Oxytocin is required for nursing but is not essential for parturition or reproductive behavior. *Proc. Natl. Acad. Sci. USA,* 93, 11699–11704.

[59] Holder, J.L. Jr, Butte, N.F., & Zinn, A.R. (2000). Profound obesity associated with a balanced translocation that disrupts the SIM1 gene. *Hum. Mol. Genet.,* 9, 101–108.

[60] Qian, W., Zhu, T., Tang, B., Yu, S., Hu, H., Sun, W., et al. (2014). Decreased circulating levels of oxytocin in obesity and newly diagnosed type 2 diabetic patients. *J. Clin. Endocrinol. Metab.,* 99, 4683−4689.

[61] Blaicher, W., Gruber, D., Bieglmayer, C., Blaicher, A.M., Knogler, W., & Huber, J.C. (1999). The role of oxytocin in relation to female sexual arousal. *Gynecol. Obstet. Invest.,* 47, 125–126.

[62] Kruger, T.H., Haake, P., Chereath, D., Knapp, W., Janssen, O.E., Exton, M.S., et al. (2003). Specificity of the neuroendocrine response to orgasm during sexual arousal in men. *J. Endocrinol.,* 177, 57–64.

[63] Anderson-Hunt, M. & Dennerstein, L. (1994). Hormones and sexual arousal: developing a method for research. In: International Women's Health Coalition - Population Council Sexuality and Gender Working Group. Learning about sexuality. New York: IWHC - Population Council (in press).

[64] Burri, A., Heinrichs, M., Schedlowski, M., & Kruger, T.H. (2008). The acute effects of intranasal oxytocin administration on endocrine and sexual function in males. *Psychoneuroendocrinology,* 33, 591–600.

[65] Fischer-Shofty, M., Levkovitz, Y., & Shamay-Tsoory, S.G. (2013). Oxytocin facilitates accurate perception of competition in men and kinship in women. *Soc. Cogn. Affect Neurosci.,* 8, 313–317.

[66] Rebelo, A. & Leal, I. (2007). Fatores de personalidade e comportamento alimentar em mulheres portuguesas com obesidade mórbida: Estudo exploratório. *Análise Psicológica,* 3, 467-477.

[67] Klockars, A., Levine, A.S., & Olszewski, P.K. (2015). Central oxytocin and food intake: focus on macronutrient-driven reward. Front Endocrinol, 6, 65. doi:10.3389/fendo.2015.00065.

[68] Lemos, I. (2005). Bulimia e anorexia: Patologias da falta e do excesso. Mental: Revista de Saúde Mental e Subjectividade da UNIPAC, 3, 81-89.

[69] Ribeiro, V.L. (2008). Obesidade e função sexual. Disponível em http://www.psicologia.com.pt.

[70] Gariepy, G., Nitka, D., & Schmitz, N. (2010). The association between obesity and anxiety disorders in the population: a systematic review and meta-analysis. *International Journal of Obesity*, 34, 407-419.

[71] Tosetto, A. & Júnior, C. (2008). Obesidade e sintomas de depressão, ansiedade e desesperança em mulheres sedentárias e não sedentárias. *Medicina Ribeirão Preto*, 41, 497-507.

[72] Scott, K., Bruffaerts, R., Simon, G., Alonso, J., Angermeyer, M., et al. (2008). Obesity and mental disorders in the general population: results from the world mental health surveys. *International Journal of Obesity*, 32, 192-200.

[73] Hunte, H. & Williams, D. (2009). The association between perceived discrimination and obesity in a population-based multiracial and multiethnic adult sample. *American Journal of Public Health*, 99, 1285-1292.

[74] Cooper, M., Shaver, P., & Collins, N. (1998). Attachment styles, emotion regulation, and adjustment in adolescence. *Journal of Personality and Social Psychology*, 74, 1380-1397.

[75] Surcinelli, P., Rossi, N., Montebarocci, O., & Baldaro, B. (2010). Adult attachment styles and psychological disease: Examining the mediating role of personality traits. *The Journal of Psychology*, 144, 523-534.

[76] Castellini, G., Mannucci, E., Mazzei, C., Lo Sauro, C., Faravelli, C., Rotella C., et al. (2010). Sexual function in obese women with and without binge eating disorder. *Journal of Sexual Medicine*, 2, 1-10.

[77] Troisi, A., Massaroni, P., & Cuzzolaro, M. (2005). Early separation anxiety and adult attachment style in women with eating disorders. *British Journal of Clinical Psychology*, 44, 89-97.

[78] Ringer, F. & Crittenden, P. (2007). Eating disorders and attachment: The effects of hidden family processes on eating disorders. *European Eating Disorders Review*, 15, 119-130.

[79] Wilkinson, L., Rowe, A., Bishop, R., & Brunstrom, J. (2010). Attachment anxiety, disinhibited eating, and body mass index in adulthood. *International Journal of Obesity,* 32, 1442-1445.

[80] Butzer, B. & Campbell, L. (2008). Adult attachment, sexual satisfaction, and relationship satisfaction: A study of married couples. *Personal Relationships,* 15, 141-154.

[81] Riley, A. (1997). Sexual behavior. In Baum A, Newman S, Weinman J, West R, McManus C. Cambridge Handbook of Psychology: Health and Medicine. (162-165). United Kingdom: Cambridge University Press.

[82] Dundon, C. & Rellini, A. (2010). More than sexual function: Predictors of sexual satisfaction in a sample of women age 40-70. *Journal of Sexual Medicine,* 7, 896-904.

[83] Araújo, A., Brito, A., Ferreira, M., Petribú, K., & Mariano, M. (2009). Modificações da qualidade de vida sexual de obesos submetidos à cirurgia de Fobi-Capella. *Revista do Colégio Brasileiro de Cirurgiões,* 36, 42-48.

[84] Kolotkin, R., Binks, M., Crosby, R., Ostbye, T., Gress, R., & Adams, T. (2006). Obesity and sexual quality of life. *Obesity,* 14, 472-479.

[85] Young, A.H. (2006). Antiglucocorticoid treatments for depression. *Aust N Z J Psychiatry,* 40, 402-405.

[86] Castro, M. & Moreira, A.C. (2003). Análise crítica do cortisol salivar na avaliação do eixo hipotálamo-hipófise- adrenal. *Arq. Bras Endocrinol. Metabol.,* 47, 358-367.

[87] Saraiva, E.M., Fortunato, J.M.S., & Gavina, C. (2005). Oscilações no cortisol na depressão, sono e vigília. *Rev. Port de Psicossomática,* 7, 1-13.

[88] Bjorntorp, P. & Rosmond, R. (2000). Obesity and cortisol. *Nutrition,* 16, 924–936.

[89] Bauer, M.E. (2005). Stress, glucocorticoids and ageing of the immune system. *Stress,* 8, 69-83.

[90] Palma, B.D., Tiba, P.A., Machado, R.B., Suchecki, D., & Tufik, S. (2007). Repercussões imunológicos dos distúrbios do sono: o eixo hipotálamo pituitária adrenal como fator modulador. *Rev. Bras. Psiquiatr,,* 29, S33-S38.

[91] Khan, S.M., Hamnvik, O.P., Brinkoetter, M., et al. (2012). Leptin as a modulator of neuroendocrine function in humans. *Yonsei Med. J.,* 53, 671–679.

[92] Champaneri, S., Xu, X., Carnethon, M.R., et al. (2013). Diurnal salivary cortisol is associated with body mass index and waist circumference: the Multiethnic Study of Atherosclerosis. *Obesity,* 21, E56–E63.

[93] Abraham, S.B., Rubino, D., Sinaii, N., Ramsey, S., & Nieman, L.K. (2013). Cortisol, obesity, and the metabolic syndrome: a cross-sectional study of obese subjects and review of the literature. *Obesity,* 21, E105– E117.

[94] Praveen, E.P., Sahoo, J.P., Kulshreshtha, B., et al. (2011). Morning cortisol is lower in obese individuals with normal glucose tolerance. *Diabetes Metab. Syndr. Obes.,* 4, 347– 352.

[95] Epel, E., Lapidus, R., McEwen, B., & Brownell, K. (2001). Stress may add bite on appetite in women: laboratory study of stress-induced cortisol and eating behavior. *Psychoneuroendocrinology,* 26, 37–49.

[96] Cavagnini, F., Croci, M., Putignano, P., Petroni, M.L., Invitti, C. (2000). Glucocorticoids and neuroendocrine function. *Int. J. Obes. Relat. Metab. Disord.,* 24, S77–S79.

[97] Figlewicz, D.P. (2003). Adiposity signals and food reward: expanding the CNS roles of insulin and leptin. *Am. J. Physiol. Regul. Integr. Comp. Physiol.,* 284, R882–R892.

[98] Golved, K.L. & van Anders, S.M. (2012). Sexual thoughts: links to testosterone and cortisol in men. *Arch. Sex Behav.,* 41, 1461-1470.

[99] Chambers, K.C. & Phoenix, C.H. (1981). Diurnal patterns of testosterone, dihydrotestosterone, estradiol, and cortisol in serum of rhesus males: relationship to sexual behavior in aging males. *Hormones and behavior,* 15, 416-426.

[100] Huerta, R., Mena, A., Malacara, J.M., de León, J.D. (1995). Symptoms at the menopausal and premenopausal years: their relationship with insulin, glucose, cortisol, FSH, prolactin, obesity and attitudes towards sexuality. *Psychoneuroendocrinology,* 20, 851-64.

INDEX

D

N

Q

R

S

saccharin, 30

sarcopenia, 19

saturated fat(s), 13, 73

scarcity, 13, 20

schizophrenia, 76

secretion, vii, x, 1, 3, 4, 5, 10, 15, 16, 30, 31, 49, 68, 69, 74, 75, 78, 79, 82

sedentary lifestyle, 13, 14, 72, 73

self-esteem, 11

self-regulation, 25

semantic information, 56, 59

sensation, 16, 70, 80

sensing, 84

sensitivity, ix, 36, 42, 44, 47, 48, 60

sensory, 4, 6, 20

Serbia, 35

serotonin, 10

serum, 7, 11, 14, 75, 88

sex, 62, 79

Sexual, v, 67, 71, 76, 77, 78, 81, 86, 87, 88

sexual activity, 82

sexual behavior, x, 68, 69, 70, 71, 72, 74, 77, 78, 79, 80, 88

sexual desire, 11, 72

sexual intercourse, 75

sexual problems, 82

sexual responses, x, 68, 70

sexuality, 44, 77, 79, 81, 85, 88

shape, 8

showing, 80

signal transduction, 49

signaling pathway, 40, 42

signals, 18, 76, 79, 88

signs, 18, 43

Sinai, 82

skeletal muscle, 18, 69, 72

skin, 5

skin diseases, 5

sleep disorders, 10, 80

smooth muscle, viii, 2, 18

SNP, 24

sociability, 22

social anxiety, 8, 60, 61

social behavio(u)r, 3, 7, 22, 37, 43, 62, 63, 75

social cognition, ix, 22, 51, 52, 55, 59, 63

social empathy, vii, 1, 3

social environment, 12, 23

social evaluation, 60

social group, 12, 59

social image, 58

social interaction(s), 4, 22, 43, 60

social memory, vii, 2, 3, 9, 12, 22, 58

social relations, 7, 23

social relationships, 7, 23

social scarcity, 20

social stress, 31

social support, 31

social withdrawal, 60

society, 3, 14, 80

sodium, 19, 44, 70, 75

solution, 30

species, 4, 6, 12, 17, 36, 41, 44, 76

spinal cord, 41, 47

Sprague-Dawley rats, 13

Spring, 81

stability, 40

state, 28, 44, 54

stem cell lines, 38

stem cells, 38, 45

stimulation, viii, x, 4, 16, 30, 35, 43, 45, 68, 70, 78, 80, 82

stimulus, 6, 7, 58, 61, 62

stress, viii, ix, x, 2, 4, 5, 6, 10, 12, 15, 17, 18, 19, 23, 26, 27, 31, 36, 37, 40, 41, 43, 45, 46, 50, 68, 70, 71, 72, 73, 75, 77, 78, 79, 82, 88

stress response, 37, 45

stressful events, 10, 11, 17

stressors, 17

stroke, 83

strong interaction, 21

structure, 6, 9, 23, 46, 84

style(s), 27, 86

substrates, 29

sucrose, 15, 30

Sun, 85

suppression, 10, 12

Y